AF651274

HORTICULTURE GUIDE FOR UTTRAKHAND

THE IMMORTAL SONG

May the sacred stream of amity flow for ever in my heart
May the universe prosper, such is my cherished desire
May my heart sing with ecstasy at the sight of the virtuous
And may my life be an offering at their feet
May my heart bleed at the sight of wretched, the cruel and the poor
And my tears of compassion flow from my eyes
May I always be there to show the path to the pathless wanderers of life
Yet if they should not hearken to me, may I bide patiently
May the spirit of goodwill enter all our hearts
May we all sing together the immortal song of humanhood

THE VALUE OF A SMILE

It costs nothing, but creates much
It enriches those who receive, without impoverishing those who give
It happens in a flash and the memory of it sometimes lasts forever
None are so rich they can get along with it and none so poor but are richer for its benefits
It creates happiness in the home, fosters goodwill in a business, and is the countersign of friends
It is rest to weary, daylight to the discouraged, sunshine to sad, and Nature's best antidote for trouble.
Yet it can not be bought, begged, borrowed or stolen, for it is something that is no earthly good to anybody till it is given away
For nobody needs a smile so much as those who have none left to give.

HORTICULTURE GUIDE FOR UTTRAKHAND

Edited by

HARBANS SINGH

M.L. DEWAN

V.R. SURI

Published by

TREES FOR LIFE

N-44 Panchshila Park
New Delhi - 110017
Phones: 6437791, 6465916

Himalaya Consortium for
Himalaya Conservation
Ranichauri - 249199
Distt. Tehri Garhwal,
Uttarakhand U.P.
Phone (01376) - 82130

All rights reserved. No part of this work may be reproduced, stored in a retrieval system, or transmitted in any form or by any means, electronic, mechanical, photocopying, microfilming, recording or otherwise, without the prior written permission of the copyright owner.

First Published 1997
© Authors

Printed by

Concept Publishing Company
A-15/16, Commercial Block, Mohan Garden,
New Delhi - 110059
Phones: 5554042, 5504042

PREFACE

"Mother Nature" has bestowed unique and rare gifts on Himalayan Region which include, Uttrakhand area. These include beautiful landscape and sparkling arrays of myriad flora in which daughters and sons of Uttrakhand region inhabit.

It is universally accepted that we owe our existance to land, landscape and vegetation thereon. However, the ecology and environment originally gifted by the Nature are fast deteriorating to the detriment of human race and other kinds of life, not only in the Himalayan region itself but also in the rest of the country and the universe.

There is great concern amongst nations regarding deterioration of environment. Temperatures are rising. If these continue to rise at this rate, people are bound to face serious problems. For instance, if the present trend continues substantial part of Maldives or even whole of it may submerge on account of the rise of sea levels. This may also affect about 15 per cent of Bangladesh and low lying areas of India, Pakistan, Sri Lanka and Vietnam.

Tree plantations and other forms of vegetation assist a great deal in reversing this trend. Horticulture crops contribute effectively in improving environment, ecology and landscape. World Resources Institute has evaluated the benefits from trees in a 50 year life-cycle. It is estimated that a tree generates 50 tonnes of oxygen during this period. Other benefits are: (i) shelter to birds at the rate of Rs. 5000 per year, squirrels and insects Rs. 5000 per year and recycling of water Rs. 6000 per year. Total benefits in tropical forests of one hactare are estimated to be Rs. 50 lakhs each year. Horticulture crop production is eminently suited for Agroclimatic-conditions prevailing in Himalayan region. Efforts have been made to develop horticulture in this region. However, there is wide gap between the technology available and actually applied in the field. This write up aims at bridging the gap effectively. It will provide easy to understand guidelines for growers, extension workers and students of horticulture sciences.

HARBANS SINGH
M.L. DEWAN
V.R. SURI

CONTENTS

THE GLORY OF TREES

O King of trees! I bow before you. Brahma is in your roots, Vishnu is in your body, Shiva is in your branches. In every one of your leaves there is a heavenly being.
Trees are like good people who care for others. They have to keep standing in the sun but they give shade to others. Whatever fruits they bear they do not eat themselves, but give them to others. How kind they are.

(*Vikrama Caritam* 65)

The whole life of these trees is to serve. With their leaves, flowers, fruits, branches, roots, shade, fragrance, sap, bark, wood, and finally even their ashes and coal, they exist for the purpose of others.

(Srimad Bhagavatam)

Trees have five sorts of kindness which are their daily sacrifice. To families they give fuel; to passers by they give shade and a resting place; to birds they give shelter; with their leaves, roots and bark they give medicines.

(*Varaha Purana* 162.42-42)

O tree! You bear fruits, leaves and flowers and protect people from scorching Sun. Whoever comes to you in scorching heat, you take away their suffering and give them coolness. This way you surrender yourself for others. That is why you are guru of all kind people. Even the most generous person cannot do things with such devotion as you. And yet show yourself to no one. Therefore, O tree, please accept my heartfelt respects.

(*Bhamini Vilasa* 89)

O tree! You are standing on the path. Live for a long time and be happy, because with your blossoms the cuckoo is happy, with your pollen the bumblebees are happy, and passers by are happy with your fruits. So live long!

(Upadesa Tarangini)

1

INTRODUCTION

The physical systems of the Himalayas, being relatively young, are highly vulnerable to the natural processes of weathering and erosion. Nature's regulatory mechanism to control the rate of this erosion is the vegetation cover. The rate of change brought about by these processes is altered by human use and misuse of the system.

Earlier, people have lived here on a sustainable basis and occupied the least vulnerable land, i.e., the flat valleys and the lower slopes of the middle Himalayan inroads in the region, it still depends at least partially, if not wholly, on itself for its basic needs (emanating from the mountainous character) and produce daily necessities like wheat, rice, timber, fruits, vegetables and live-stock. The region further sustains a vast tourist industry, partly due to the presence of places of pilgrimage—Haridwar, Rishikesh, Yamnotri, Gangotri, Kedarnath, Badrinath and Hemkund etc.

Trees for Life - (TFL)

Trees for Life was founded as a World Organization by Balbir Mathur, with a Trust registered at Allahabad. The Himalayan Operations are directed by Dr. M.L. Dewan, Advisor, TFL, and Founder Chairman of HIMCON (Himalayan Consortium for Himalayan Conservation).

Trees for Life provides management and know-how to people to plant and care for food-bearing trees to fight hunger and malnutrition. Trees for Life is a non-profit organization.

TFL was started with the following objectives:

(i) To provide techniques and materials required to plant, cultivate and harvest food bearing and other trees and to undertake plantation and/or assist others in plantation of such trees.

(ii) To carry out programmes and talks at public places, schools and service clubs etc. to provide education, training, infusion of appropriate technology and awareness amongst people for plantation of trees to fight malnourishment, hunger and for improvement of ecosystems.

(iii) To develop a series of clear, concise and candid materials for the purpose of being made available to all such public organisations and societies who are genuinely concerned with the problem of fighting hunger and malnourishment, so as to enable them to convey the message and necessity of planting more food bearing and other trees in the most correct manner.

Himalayan Consortium for Himalayan Conservation (HIMCON)

Focusing on the fragile eco-system of the Central Himalayas, HIMCON, in conjunction with TFL, and other voluntary and people's organisations is attempting to find alternative development approaches to bring about a qualitative change in the lives of the simple communities through their own active involvement and participation. Thus HIMCON as partner of TFL in the Himalayas enlarging the scope and activities of TFL in the central and western Himalayas, is following the strategy well established for horticulture growth and development of the region.

Horticulture Guide for Uttrakhand (U.P.)

One of the activities of TFL/HIMCON is to bring out a Horticulture Guide for Uttrakhand (U.P.) under the able leadership of Dr. Harbans Singh a well known Horticulturist who has worked as Horticulture Commissioner, Government of India and has been the person behind Horticulture Revolution in Himachal Pradesh. He has been ably assisted by Mr. V.R. Suri, a long-term associate of Dr. Harbans Singh.

I trust this Guide will be of help to many people, NGO's, People Organization and even Government officials in Uttrakhand and other Himalayan regions.

Any suggestions or additional information will be welcome for the next edition.

Please write to:

Dr. M.L. Dewan
TFL/HIMCON
Village & PO.
Ranichauri, 249199
Distt. Tehri Garhwal (U.P.)

2

UTTRAKHAND

Uttrakhand area of the Uttar Pradesh State covers eight districts, three in the Kumaon Division viz. Almora, Pithoragarh and Nainital; five in the Garhwal Division viz. Uttarkashi, Chamoli, Tehri, Dehradun and Pauri Garhwal. The inner Himalayas adjoin the plain areas, the middle Himalayas consist of folded mountains and valleys ranging upto a height of 2500 m above sea level and the outer Himalayas consist of alpine meadows and snow clad ridges and high peaks like Nanda Devi, Trishul, Kedar Nath, Neel Kanth, Chaukhambha and Panchchuli.

Uttrakhand area lies between 28°-32° Latitude and 77°-91° Longitude. Relevent details of the 8 districts are as under:

Latitude and Longitude of U.P. Hill Districts

District	Latitude	Longitude
Almora	29°-38°	79°-81°
Pithoragarh	29°-30°	79°-81°
Nainital	28°-30°	78°-81°
Uttarkashi	30°-32°	78°-79°
Chamoli	30°-31°	79°-80°
Tehri	30°-31°	78°-79°
Dehradun	29°57'-31°2'	77°35'-78°29'
Pauri	29°20'-30°15'	78°10'-79°10'

Two of the eight districts viz. Dehradun and Nainital cove[r] large plain areas and are relatively more developed than th[e] other six districts. The Uttrakhand region covers an area of 51,125 sq. km. (17.4 per cent of the total state area) with a population of 59.26 lakh (1991 census) which is 4.3 per cent of the State population; thus while the State density of population per sq. km is 472, in the Uttrakhand region it is 116 person per sq. km. Among the 8 districts of the region highest density per sq. km was 332 in Dehradun and lowest, 30 in Uttarkashi district as will be seen in the following table:

Area, Population, Density

District	Area (sq. km)	Population (000s) 1991	Density per sq. km.
Almora	5385	837	155
Pithoragarh	8856	566	64
Nainital	6794	1540	227
Uttarkashi	8816	240	30
Chamoli	9126	455	50
Tehri	4421	500	131
Dehradun	3088	1026	332
Pauri	5438	682	126

The scheduled caste population of the region is 16.7 per cent (State percentage of 21.0). Rural population constituted 78.3 per cent population. Uttrakhand region reported higher literacy rate in 1991 census-59.6 per cent against the State's average of 41.6 per cent.

According to 1991 census workers constituted 41.9 per cent of the total population, majority of them being cultivators (58.11% of the main workers). What is significant is that female workers constituted 41.0 per cent of the total workers as against the State average of 17.0 per cent. This, by and large, is explained by the high level of out-migration of males. The state of women workers in U.P. Hills was commented upon by the Working Group on Hill Area Development for the Seventh Plan thus: "The workload of the hill women and the drudgery involved is extraordinary. She works 14 to 16 hours a day spending as much as 2 to 5 hours for collection of fodder and anywhere from 4 to 7 hours for collection of firewood (2-3 days in a week). In addition, there is the chore of fetching water and in many villages they grind their wheat themselves because no power *chakki* is available within walking distance.

The average rainfall in Uttrakhand hills is 1525 mm per year, the same ranging between 2540 mm in Nainital to 1016 mm in Pauri as can be seen in the following table.

Average Annual Rainfall in Uttrakhand

Districts	Rainfall in mm
Almora	1324
Pithoragarh	1820
Nainital	2540
Uttarkashi	1524
Chamoli	1100
Tehri	1043
Dehradun	2360
Pauri	1016
Av. for the Region	1525

Of the total area of the region a high percentage of 63.9 was reported to be under forests and only 12.5 per cent of area classified as net sown. The remainder constituted not available for cultivation (6.8%), unfit for cultivation (5.5%), fallow and other cultivable waste (11.3%). Between the 8 districts of the region land classified under forests ranged from 51.8 per cent in Pithoragarh to 88.6 per cent in Uttarkashi district. The net sown area percentage was lowest in Uttarkashi (3.6%) and highest in Nainital (29.5%).

Irrigation facilities in the Uttrakhand area were at a much lower scale as compared to the whole of Uttar Pradesh. Of the net sown area only about 25 per cent area was irrigated while the corresponding figure for the State was 61 per cent. As between the 8 districts of Uttrakhand, Nainital district had 81 per cent of its net sown area under irrigation (canals & pumpsets). In Dehradun district corresponding figures were 49 per cent. Very low figures were reported from Chamoli (5.4%), Pithoragarh (8.1%) and Pauri (8.9%).

Land holdings in the Uttrakhand area were predominantly marginal and small, more or less on the same pattern as for the State as a whole.

Holding	Percentage Distribution	
	No. of holding	Area under the holding
Below 1 ha	70.9	25.6
1-2 ha	16.8	24.3
2-3 ha	6.3	15.7
3 ha and above	6.0	34.4

Area under different crops in the Uttrakhand area is shown below:

Crop	Area under the Crop	
	Irrigated	Total
Paddy	1,50,923	2,59,440
Maize	631	29,162
Ragi	138	1,39,315
Wheat	1,62,079	3,58,378
Barley	1,080	27,686
Other Cereals	93	92,732
Pulses	1,979	42,851
Fruits & Vegetables	6,988	29,831
Oilseeds	5,784	11,687
Others	59,453	1,00,293
Total Cropped Area	3,89,148	10,91,375

The two basic infrastructural facilities are the *pucca* roads and electricity for various uses. In Uttrakhand the length of *pucca* roads for 1000 sq. km. area was 227.1 km. As between the different districts, Tehri Garhwal had 469.4 km *pucca* roads for 1000 sq. km area, Dehradun district 430.7 km while in Uttarkashi district this figure was only 87.2 km. A little over 73 per cent villages in Uttrakhand were electrified, this percentage being 100 per cent in Nainital district and 59.5 per cent in Pauri Garhwal. The per capita consumption of electricity in the Uttrakhand area was 217 kW. against the overall state average of 170 kW. Among the 8 districts, the highest per capita consumption was reported in Dehradun (500 kW) and lowest in Chamoli (32 kW).

The Agro-climatic Regional Planning Project's delineation of different agro-climatic regions of the country places Uttrakhand eight districts as sub-zone 3 in the Himalayan Region. The other two sub-zones in the region are Jammu and Kashmir State (Sub-zone 1) and Himachal Pradesh (sub-zone 2). These three Himalayan sub-zones in respect of some of the basic characteristics stand as under:

Important Characteristics—Hill Region—North West Himalaya

Sub-zone	Area sq. km.	Population 1991 census (lakhs)	Density sq. km	Literacy (per cent)	Work Participation (per cent)		
					Male	Female	Total
1. Jammu & Kashmir	2,22,240	77.2	NA	NA	NA	NA	NA
2. Himachal Pradesh	55,670	51.7	93	63.0	50.6	34.8	42.8
3. Eight Hill districts of U.P.	51,124	59.3	116	59.6	48.7	35.2	41.9

N.A. : Data not available.

Av. Rain-fall (mm)	Percentage of total area			Crop-ing Inten-sity	Net irriga-ted percen-tage of net sown area	Percent-age of villages electri-fied	Length of Pucca Road per 1000 sq. km
	Forest	Not-available for culti-vation	Net sown area				
1003	61.0	5.7	16.2	141	40.8	N.A.	N.A.
1360	29.5	6.5	21.5	164	17.0	N.A.	N.A.
1525	63.9	6.8	12.5	164	24.9	73.5	227.1

Status of Horticulture in U.P. (Hill) and UP (Plains)

Data of area and production of fruits, vegetables and potatoes in U.P. (Hill) and UP (Plains) as per the State Government agencies is given in the attached two tables. It will be seen there that while U.P. hills comprise 17.4 per cent of the total state area U.P. plains account for 82.6 per cent of the said area. The relative portion of area under fruits, vegetables and potatoes in the two tables shows following distribution.

Percentage of Area under Fruits, Vegetables Potatoes - 1992-93

Region	Percentage of area under		
	Fruits	Vegetables	Potatoes
U.P. Hills	19.0	6.2	4.7
U.P. Plains	81.0	93.8	95.3

While the area under fruits is relatively higher in U.P. hills, those under vegetables and potatoes the same is much lower. This may be due to the marketing problems typical of the hills.

In terms of productivity the performance of U.P. hills is much lower in respect of Fruits 2.62 tonnes per ha. as against 8.81 tonnes per ha. in U.P. plains, Vegetables 4.99 tonnes per ha. in U.P. hills against 13.49 tonnes in U.P. plains.

In Potatoes, however productivity per ha. was higher in U.P. hills (20.63 tonnes) against 18.78 tonnes of U.P. plain. But the situation under fruits and other vegetables calls for increased technological back-up, concerted extension activity and creation of marketing infrastructure.

Area, Production and Productivity of Fruits and Vegetables including Potatoes in Uttrakhand

Sl. No.	Particulars	Unit	Years 1990-91	1991-92	1992-93
1.	**Area**				
	(a) Fruits	Hactares	169830	173055	176170
	(b) Vegetables	Hactares	53050	57065	61072
	(c) Potatoes	Hactares	16510	17515	18516
	Total		239390	247635	255758
2.	**Production**				
	(a) Fruits	Tonnes	440200	452000	461793
	(b) Vegetables	-do-	264343	276400	305000
	(c) Potatoes	-do-	322300	431200	382000
	Total		1026843	1069600	1148793
3.	**Productivity**				
	(a) Fruits	Tonnes/ha.	2.59	2.61	2.62
	(b) Vegetables	-do-	4.98	4.84*	4.99
	(c) Potatoes	-do-	19.52	24.61	20.63
	Overall Productivity		4.29	4.32	4.49

Source: Directorate of Horticulture & Food Processing, Ranikhet, Uttar Pradesh (Hills)

* Potatoes productivity is even higher than in plains which is doubtful.

Area, Production and Productivity of Fruits and Vegetables including Potatoes in Uttar Pradesh (Plains)

S. No.	Particulars	Unit	Years 1990-91	1991-92	1992-93
1.	**Area**				
	(a) Fruits	lakh ha.	7.49	7.65	7.80
	(b) Vegetables	-do-	9.21	9.25	9.50
	(c) Potatoes	-do-	3.62	3.70	3.80
	Total		20.32	20.60	21.10
2.	**Production**				
	(a) Fruits	Lakh tonnes	66.00	68.77	72.46
	(b) Vegetables	-do-	124.27	132.00	142.50
	(c) Potatoes	-do-	68.00	80.66	85.50
	Total		258.27	281.43	300.46
3.	**Productivity**				
	(a) Fruits	Tonnes/ha.	8.81	8.99	9.29
	(b) Vegetables	-do-	13.49	14.27	15.00
	(c) Potatoes	-do-	18.78	21.80	22.50
	Overall Productivity		12.71	13.66	14.24

Source: Performance Budget of Department of Horticulture & Fruit Utilization Uttar Pradesh (Plains) 1994-95.

3

ROLE OF HORTICULTURE CROPS IN HUMAN DIET AND LIFE

The importance of organic living needs no emphasis now. The global research work has established a significant link between diet, health and old age. A diet rich in fruits and vegetables is associated with lower incidence of diseases, like cancer, heart attack, etc. It is recognised that vegetarians live longer as they are healthier and less prone to high blood pressure, heart attack and cancer.

The scientists at Centre for Sciences in Public Interest, Washington DC have developed a handy rating system for different foods titled "Nutrition Score Card—Your guide to better eating."

The system assigns a simple numerical value on the scale of plus 200 to minus 100. In the unique system less nutritious "empty calorie" food gets negative scores indicating their potential for harm under this system (Jacobsons). Each food is rated according to its contents of vitamin A, B1, B2, C, iron, calcium, protein, fibre, trace minerals and unsaturated fats or added sugar.

The score card for some of the foods is as follows.

Score Card for Food Nutrition Score Board*

S.No.	Foods	Score		
1.	Cantaloupe	99		
2.	Water melon	74	Chicken liver	158

(Contd.)

S.No.	Foods	Score		
4.	Peach	29	Tuna fish	55
5.	Banana	26		
6.	Apple	12		
			Natural foods better than processed foods.	
7.	Broccoli	116		
8.	Kala	118	Almonds,walnuts, cashews,	
9.	Turnip greens	111	raisins and even popcorn have	
10.	Collard greens	126	positive value. Cookies and candy	
11.	Celery	7	have minus value and ratings	
12.	Cucumbers	8		
13.	Iceberg lettuce	10		
	Beverages			
14.	Orange Juice	62		
15.	Skim milk	49		
16.	Whole milk	39		
17.	Tomato Juice	37		
18.	Coffee	0		
19.	Coffee with sugar	-12		
20.	Soft drink	-92		

* Centre for Science in Public Interest.

The importance and value of the fruits and vegetables is regularly being emphasised in large number of publications including news papers for public awareness and use. Some of the important headlines are here reproduced. These are just a small sample of the huge material.

There is more to tomato than colour
Curative value of pea
Therapeutic value of (cashew) almonds and walnuts
Vegetarian food makes one brave.
Eat your veggies, and live longer

Iron-rich spinach is more loaded raw
Amla is packed with vitamin C
Tea and apples are good for heart
Copper makes water germfree
Methi for birth control
Papaya—the wonder fruit
Medicinal virtue of lemon
Green bananas cure ulcers
Vegetables help prevent cancer
Banana for virility.

4

FUNCTION AND ROLE OF PLANTS IN NATURE

The plants play a fundamental role in the life and survival of human race on earth. These are Nature's chemical industry. Each plant (or species) work as an independent factory which directly or indirectly support life in the universe. But for plants, life on earth would have perished long ago. Horticulture species are important part of plant kingdom. Besides the functions mentioned above these provide indispensable and valuable food, nutrition, protection and prevention from old age and diseases and a large variety of herbs (Medicinal Plants).

The functioning of the plants as nature's chemical factory is very complex and interesting. Many manufacturing functions are common to all plants but there are variations amongst the species and each species may be producing only specific chemicals.

Plant leaves are very important component of plant laboratories Photosynthesis carried out by leaves with sunlight results in production of eleborate material including chemicals and medicines. But this requires support of other parts of the plant for providing raw material for the ultimate product. These raw materials are supplied by the roots through stem. A rough indication of the working of the system of the plants (including its different parts) as a flawless, well coordinated and integrated factory is indicated in the diagram.

FIGURE

The Plant as a Factory

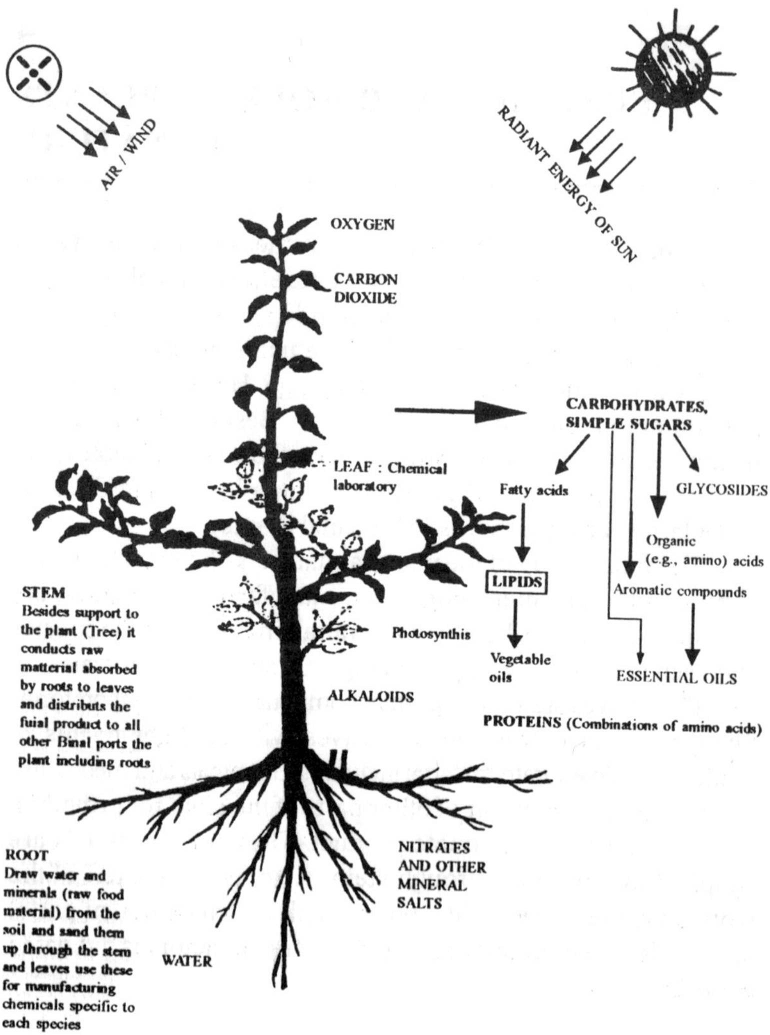

5

PROBLEMS, CONSTRAINTS AND INADEQUACIES

The horticulture industry, as in other parts of the country, in Uttrakhand also, is facing host of problems and inadequacies. Undulating topography, varying climate and soil conditions, scanty culturable land, serious soil erosion, etc. are some of the over-all problems relevent to all the sectors of the economy. The State Plan document concedes that the Uttrakhand area recorded a lower growth of 2.4 per cent (Commodity producing sector) during 1985-90 against the 3.8 per cent growth for the whole of the State. The document further states that Uttrakhand "shares common problems with other special category hill States like Himachal, J&K, etc. but Uttrakhand is getting relatively low per capita central assistance."

The different problems faced by the horticulture industry of the area are summarized under respective sub-sectors.

A. Production Constraints

(i) Low production per unit area
(ii) Fluctuating or unstable production from year to year
(iii) Low quality produce
(iv) Low financial returns per production unit.

There are many factors which contribute to this situation. Important of these are:

(i) Inadequacy of technologies specially suited to the area.

(ii) Poor quality planting material including seeds and rootstocks.
(iii) Wide gap in technology available and actual application in the field.
(iv) Low density plantation, gaps and wastage of space.
(v) Poor canopy and tree management.
(vi) Water use inefficiency in irrigated areas.
(vii) Moisture stress at critical periods of plant and fruit growth particularly under rainfed conditions.
(viii) Inadequate plant nutrition and organic matter.
(ix) Lack of integrated pest control.
(x) Inadequate credit.
(xi) Faulty and inadequate data base.
(xii) Absence of appropriate tools, implements and machinery and shortage of skilled labour.
(xiii) Menace of wild animals and stray cattle in some areas.

B. Post Harvest Handling and Marketing Constraints

(i) Absence of worthwhile markets in the area.
(ii) Wide gap in technology available and applied.
(iii) Post harvest quality control almost non-existent resulting in considerable wastage and damage.
(iv) A part of the produce handling and marketing is in the hands of the middlemen who do not bother about maturity standards, grading and scientific packing.
(v) Absence of pre-cooling and cold storages (for other than potato).
(vi) The marketing system is more middlemen than producer and consumer oriented.
(vii) Long distance from major consumption areas and rough, rugged and poorly maintained roads.
(viii) Inefficient and inadequate facilities in the market.
(ix) Lack of market intelligence and market information service.

C. Processing

(i) Absence of processing (including drying facilities).
(ii) Poor consumption base and sales of the product.

(iii) High duties and taxes.
(iv) Non-availability of raw materials specially suited to the industry.
(v) Absence of linkage between the processor and grower.
(vi) High cost of basic packing material and other inputs.
(vii) Lack of any incentive for the use of horticulture produce in processing as against synthetic products which are cheap.
(viii) High transport rates for plain markets.
(ix) Inadequate sales and publicity campaigns.
(x) A totally disorganised sector.

D. Organisation and Management

(i) Inadequate staff as a whole and lack of professionalism.
(ii) Lack of professionalisation of the set up.
(iii) Absence or inadequacy of orientation and training in new technologies.
(iv) Inadequate extension services.

E. Credit and Investment

(i) Inadequate or mostly non-availability of credit.
(ii) Complicated procedure and rules.
(iii) Lack of short-term loans for orchard management and marketing.
(iv) Poor government investment on horticulture.

F. General

(i) Absence of long range planning for fruit and other perennial crops.
(ii) Inadequate (or non-existent) linkages between production, marketing and processing.
(iii) Inadequate research support particularly location specific field oriented problems.
(iv) Poor linkage between research and extension.
(v) Inadequate or lack of assessment of successful stories in the field of production, post harvest handling, marketing

and processing.

(vi) Poor data base and some times wrong figures.

(vii) Inadequate and insufficient research support and linkages.

(viii) Absence of policy document on the promotion of horticulture crops.

6

HORTICULTURE TECHNOLOGY

There are large variations in the climatic conditions of the Uttrakhand region (UKR). The rainfall ranges from 1016 mm to 2540 mm annually in the different districts. The mountain height range from 400 metres to around 7000 metres. There are low height areas between high mountains. Several valleys have been carved out of mountains due to the impact of rivers and their tributaries. The topography is broken and there is no uniform pattern for the same. It is, therefore, not possible to demarcate horticulture zones in continuous or regular basis on a map. This is what characterises most of the hill regions of India.

The Uttrakhand Region can broadly be divided into 3 specific groups from horticulture production point of view. The zones and the fruit crops that can be grown are:

Zones	Most significant fruit crops
1. High Hill Zone including dry areas. (generally about 1500 metres and above (ASL)	Apple, Pear, Cherry, Walnut Hazelnuts, Almonds
2. Mid Hill Zone (910 to 1500 metres)	Peach, Plum, Apricot, Almond, Walnuts, Pecanuts, Olive, etc.
3. Low Hill Zones (upto 910 metre ASL) including Tarai and Valley areas	Sub tropical fruits like Mango, Litchi, Citrus, Guava, Peach, Plum, etc.

The above distribution is neither exhaustive nor rigid. These are only broad outlines. Climate, particularly temperatures are governed by several factors such as height above sea level, nearness to perpetual snow line, nearness or away from the plains, aspect of the hill, etc. Thus valley in the interior (say Chamoli district) at lower elevation may have a cooler climate than higher hills on the outerside or near the plains. At the same height northern part of a hill range will be cooler than south or south-western parts. All these aspects plus the availability of moisture play a crucial role in determining the kinds of trees that can be grown successfully in each zone. Several of the crops mentioned above can be grown in more than one zone. The local climatic conditions at the site must be taken into consideration at the time of selection of kinds and varieties.

It must be remembered that we cannot succeed in commercial growing of fruits where the nature does not intend to grow them.

A. Fruit Crops (Horticulture Tree Crops)

System of Plantation

By and large square system of planting fruit trees is generally followed in plains and same copied in the hills. However, this system does not fit in the land scape of the hills which is slopey and of broken topography. The contour system of plantations which fits in the natural topography of the hills has been found to be most suitable under these conditions. Since most of the plantations are unirrigated (or rainfed) special emphasis is required to be laid on the conservation of soil and moisture *in situ*. This is being explained under rainfed horticulture.

Normal space recommended under the conventional system and trees per hectare are given below:

Normal Distance Recommended

Fruit	Distance (Metre)	No. of Trees (Per ha)
Apple/Pear	6×6	277
Mango/Litchi	9×9	120
Citrus	6×6	277
Guava	6.5×6.5	236

Peach/Plum	6 × 6	236
Cherry	6 × 6	277
Papaya	3 × 3	1089
Walnut	8 × 8	156
Almond	5 × 5	400

Modified planting distance recommended of trees per ha.

Fruit	**Distance (Metre)**	**No. of Trees (Per ha.)**
Apple/Pear	9 × 5	222
	9 × 6	176
Mango/Litchi	12.5 × 6.5	120
Citrus	5 × 8	333
Guava	4 × 6	416
	5 × 6	333
Peach/Plum	4 × 6	416
Cherry	4 × 6	416
Papaya	2 × 2	2500
Walnut	6 × 8	210
Almond	4 × 5	500

The distances are indicative and may depend upon kind, variety, rootstock, soil, its fertility and slope.

Contour Plantations

This should be generally followed on hill slopes along with moisture and soil conservation *in situ* practice. This system is more or less a rectangular pattern as against square system. In this sytem total space for each kind is almost or nearabout the same as mentioned above. That is, if the space in case of apple and pear is 6 × 6 metre, the total space per tree comes to 36.0 sq. metre. In case of contour plantation line is drawn along contour (on the slope). The contours below and above this line may not be at the distance indicated in the table. It can either be 7.5 metre or 8 metre or even 6.5 metres at many places depending upon the degree of slope. Nothing to be worried. The distance between the trees on the contour line is adjusted depending upon the distance between 2 contours. This can be closer or wider. It may be useful to have little closer spacing as tree canopy on different contours will not clash or join together like the plantation on plain areas.

Important Varieties and Pollinizers of Temperate Fruit Crops

S.No.	Fruit Crops	Varieties	Pollinizers Varieties
1.	Apple		
	(i) Early	Early shan burry, Benoni Fanny, Red June, Vens Delicious, Tydeman Early Worcestor.	Early Shan burry, Benoni, Fanny, Tydeman Early Worcestor.
	(ii) Mid	Red Delicious, Royal delicious, Red Gold, Winter Banana, Golden Delicious Red Spur Delicious, Gold Top Red, Oregon Spur, Red Chief, Star Spur, Red Summer, Red, Starking Delicious, Lord Lambourne, Chaubattia Princess.	Red Gold, Golden Delicious, Lord Lambourne
	(iii) Late	Buckingham, Rymer, Ambri Granny Smith White Red.	Buckingham, Rymer, Granny Smith.
2.	Pear		
	(i) Early	Thumb Pear, Dr. Jules Goyat	
	(ii) Mid	Beurre-De-Amanalis, Doyenne-Du-Commice, Victoria, Conference.	
	(iii) Late	Winter Nelis, Beure Hardy Max red Barlet, Jargoualle.	Winter Nelis, Beurre Hardy, Bartlet.
	(iv) For Valley area	China pear, Keiffer, Gola Smith Leucounte.	China pear, Gola.
3.	Cherry		
	(i) Early	Black Heart, Alton, Stella	Black Heart, Alton
	(ii) Late	Governer Wood, Fragmore Early.	Governer Wood, Fragmore Early.

S.No.	Fruit Crops	Varieties	Pollinizers Varieties
4.	Peach		
(i)	Early	Craw Ford Early, Early Elberta Starking Delicious, Early White Giant.	
(ii)	Late	Alexander, Perigrine, July Elberta.	
5.	Plum	First Plum, New Plum, Santarosa, Mariposa, Maynard.	
6.	Almond	I.X.L., California Paper shell, Ne Plus Ultra, Drake, Briggs, Hard shell, Chaubattia Local Super Nova.	Drake, I.X.L., Ne Plus Ultra
7.	Walnut	Tutle, Xinea, Hartley, Japan Walnut, Eureka Lake English Lake Govind, Placentea, Poyne, Tehama.	
8.	Pecan Nut	Mahan, Nelis, Barket, Discovery, Stuart curtis cape Fear	
9.	Hazel Nut	Ronda Roman, Royal, Romische Zeller, Daviana, Cosford Lambart.	
10.	Chest Nut	Italian, Japanese, Spanish Lady, Nainking.	
11.	Apricot	Charmaghz, Turkey, Kaisa, Chaubattia Madhu, Moor park, New Large Early, Royal, Safeda, Australian, A. Erani	

Soil Requirements of Fruits

Although the fruit trees exhibit a wide range of adaptabilities, their performance will be at their best if the optimum soil conditions required for different kinds or species of fruit are provided.

Water-logged, poorly drained, very shallow, soils with excessive salts should always be avoided. A deep loam or sandy loam soil is good for many fruits. Trees grown in shallow soils are more likely to be alternate bearing than those in deep and well aerated soils.

Good drainage of soil is one of the important factors in successful fruit cultivation. Well drained soil contains almost no free water upto 4 feet except after rain. Imperfectly drained soils will have ground water at about 2 feet for a month or so while most poorly drained soils will be water-logged within one foot.

The optimum requirements of soils for different kinds of fruits are given below:

S.No.	Name of the Fruit	Soil Requirements
1.	Apple	Deep, rich loam, limy and retentive of moisture but well drained. The sub-soil should also be deep and well drained.
2.	Pear	This is a very hardy fruit tree and can tolerate the extremes of soil and soil moisture conditions. Although it responds to a rich and good soil it can be grown in shallow soils, or clayey sub-soil and moist soils. It can also resist a fair amount of alkali soils in the soil. In fact it can thrive well under the conditions where most other trees will fail.
3.	Apricot, Peach and Plum	The cultivation of these fruits can be adjusted by the use of the root stock suitable for different set of soils conditions. Peach stock requires light, warm and sandy loam soil. It can also be grown in the soils having gravels. Apricot stock grows well in slightly heavier and moist soils than peach.

S.No.	Name of the Fruit	Soil Requirements
		The plum on the other hand can tolerate much heavier, stiff and moist soil. For heavy soils and sub-soils which are too retentive of soil moisture plum stock should be used.
4.	Almond	Although almond can also be budded on apricot, plum, peach and almond, it does best on the last two stocks. The roots of almond go deep and require dry, and warm soils and sub-soils. Poorly drained, water-logged soils should always be avoided. It can withstand considerable drought.
5.	Persimmon	Sand loam, rich and well drained soils are most suitable for its cultivation.
6.	Cherry	Deep, loamy alluvial and well drained soils. Standing water is very injurious to this tree.
7.	Walnut	Deep, rich, well drained but retentive of moisture and free from impervious layer in the sub-soil.
8.	Fig.	Well drained loamy soils.
9.	Pomegranate	Deep calcarious soil
10.	Grape	Light sandy loam is the best. It can however, grow in any kind of soil that does not have much water.
11.	Citrus fruits	Do best in well drained, deep and loamy soils.
12.	Mango & Jaman	Deep light loamy and well drained soils are most suitable for these fruits.
13.	Banana	Rich, moist, and well drained soil
14.	Loquat	Deep light loam and well drained soil

S.No.	Name of the Fruit	Soil Requirements
15.	Guava	Deep, sandy loam, can tolerate varying degree of soil moisture conditions quite tolerant to alkalinity
16.	Olive	Can thrive on soils which are dry and too rocky for other fruits.
17.	Papaya	Light, sandy loam and well drained.

The type of soil mentioned against different kinds of fruits indicate their optimum requirements under which they will yield maximum possible production per unit area. It does not, however, mean that the fruits will not grow if there are some deviations in soil conditions. As already said earlier the fruit trees have wide range of adaptabilities and commercial orchards can also be made successful in other types of soils excepting in cases of very defective soils mentioned earlier. The texture and usefulness of the soils to the particular kind of trees can also be adjusted to some extent by judicious soil management practices in the orchard.

Planting Seasons

There are two distinct categories of trees, evergreen and decidous (which shed their leaves in winter). Evergreen can be planted in monsoon and spring seasons. If there is shortage of water for irrigation, spring plantation should be avoided as this is followed by high temperatures and dry conditions in summer. In such cases monsoon planting should be prefered. Decidous trees are only planted in winter (December-January and February) when they have shed their leaves. The planting seasons in lower elevations where spring starts early ends in end of January or beginning of February. It depends upon local climatic conditions.

In snow bound areas the planting is recommended either before snow fall or on its melting. The branches if any of the young plants are likely to break with heavy snow fall if planted before snow fall.

Preparation for Planting

Digging of pits of 1 × 1 metre and one metre depth is recommended in the soils which are either clayey or have stones and boulder. This exercise is not necessary if the soils and sub-soils upto a depth of one metre are deep alluvial or loamy. The pits after exposing to sun for a couple of weeks are filled with the mixture of top soil and farm yard manure or compost in the ratios of 50:50 or 2:1 depending upon the local situation. The pits should be filled well in advance so that the soil settles down before planting. The soil level of the pits at the time of planting should be the same as that of the field. Transplanting on contour can also be done the same way.

Pollination

The pollination arrangements are of critical importance in several temperate zone and other plants. Adequate pollinators must be provided for (if required) at the time of plantation. Many varieties are self-unfruitful though their pollens are viable. The pollinating variety besides being compatible with the variety to be pollinated must also be flowering at the same time.

Pollination Plans

Depending upon the varieties the plans can be 11, 15, 25 or 33 per cent pollinators. If the pollinator varieties are good commercially any plan can be taken up. However, if it has low-commercial value in the market, the pollinators should be minimum. In case of apple golden delicious is a good pollinator for starking delicious and 11 per cent will suffice.

Rough pollinating plans are shown in next two pages. (Pollination Plans)

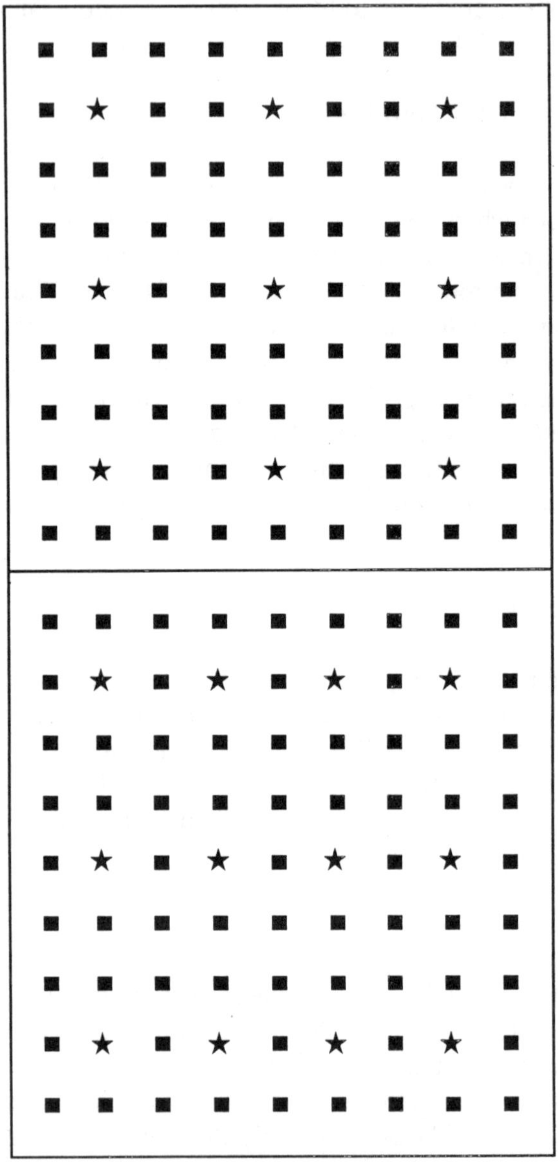

11% Pollinizers
Third plant in every third row.

15% Pollinizers
Every alternate plant in third row.

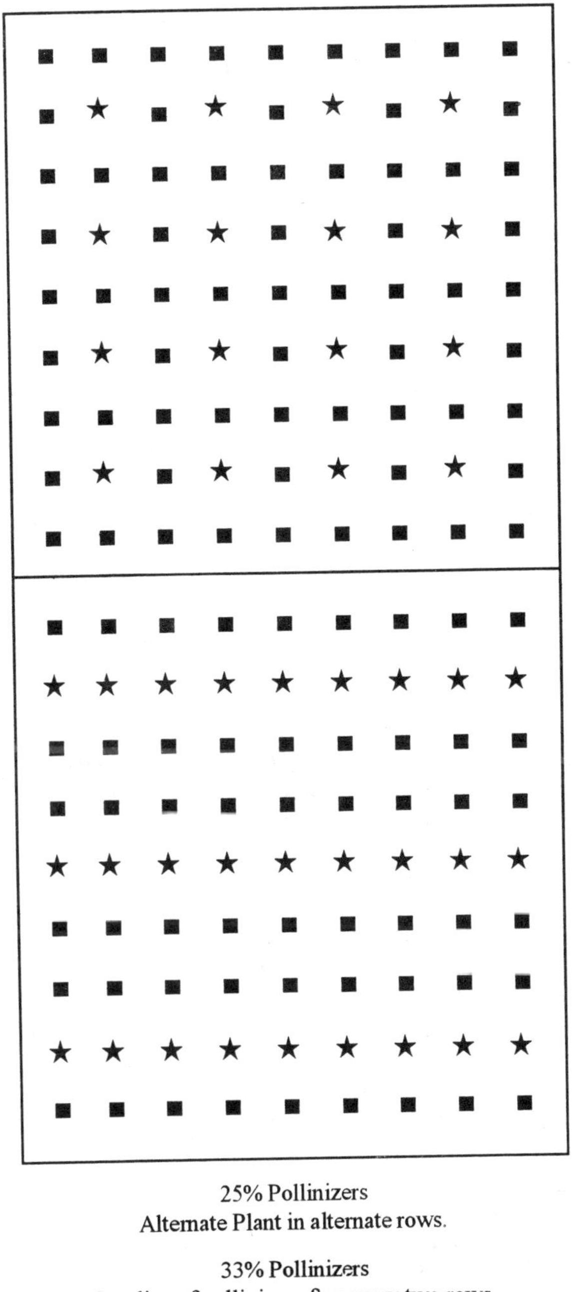

25% Pollinizers
Alternate Plant in alternate rows.

33% Pollinizers
One line of pollinizer after every two rows.

IMPORTANT BEE FLORA

Sl. No.	Particulars	Suitable for all elevation zones	Suitable for High hills	Suitable for mid hills	Suitable for low hills
1	2	3	4	5	6
1.	Roadside, avenue, wind-break and other trees	Eucalyptus sp., Willow, 'Kainth' (Wild Pear)	'Daral' (*Cedrala Serrata*). Robinia, Wild Cherry, Wild peach, Wild plum, Wild apricot	'Tun' 'Amaltas' (*Casisa-fistula*). 'Pride of India' (*Lagrestomia*), 'Puna' (*Ehretia* acuminata), 'Soapnut' 'Biul' (*Grewia opositifolia*), Wild peach, Wild plum, Wild apricot, Bohenia sp.	'Arjun' 'Barna' (*Crataeva* sp.). 'Amaltas' 'Mosquite' (*Prosophis* sp.) 'Neem' Pride of India 'Puna' 'Shisham' 'Siris' 'Jaman' 'Biul' (Bohinia sp.)
2.	Bushes, shrubs and creepers	Roses Not recommended as dense plantations, (hedges etc. for apple, pear and stone fruit orchards).	'Kshmal' (*Berberris* sp.) Plectranthus sp., 'Bhekhal' (*Rubus* sp.).	'Kshmal' (*Plectranthus* sp.), 'Bhekhal'.	Coral creeper
3.	Clovers and other crops	Sweet clover, Specie Lucerne (alfa alfa), Maize	Buckwheat White clover Red clover Pink clover	'Sarson' 'Toria'	'Berseem' 'Sarson' 'Shaftal' 'Toria'
4.	Annual ornamentals	Ageratum, Aster, Buttercup, Calendula, Coreopsis,			

1	2	3	4	5	6
		Cornflower, Cosmos, Gaitardia, Goldrod, Holly Hock, Hydrangia, Nasturtium, Poppy, Portulaca, Sunflower, Violet, Zinia.			
5.	Vegetables	Carrot, Corriander, Cabbage, Cauliflower, Knolkhol, Mongra, Raddish, Turnip, Bitter gourd, Bottle gourd, Cucumber, Pumpkin, Garlic, Lady finger, Onion.			

Pollinating Agents

Cross pollination is done by air mostly in walnuts, chestnuts, peacanuts, pistachionuts where pollens are profusely produced and carried by wind to long distances. In some cases it may be over a kilometre. Honey bee is the most important pollen carrier in case of apple, cherry, plums apricots and peaches. Its role is vital in case of cross pollinated crops. In such cases good fruit set largely depends upon the activities of insects particularly bees.

Top Working of Pollinizer Variety on the Existing Trees

Some orchardists commit the mistake of planting self-unfruitful varieties in a block. In some other cases either the distribution of pollinizer is not well planned or these are totally inadequate. All such orchardists should make good the pollinizer requirement by grafting or budding a few branches with pollinizer variety on each tree or by topworking suitable, well placed whole trees. In the former cases generally two branches on the periphery of tree are grafted, these should be on the higher side and situated in opposite direction. Thinning of fruits on the pollinizer branches should be done early to secure good annual flowering. In the latter case all the main branches of the tree are grafted over and the whole tree is changed to a pollinizer variety. One of the plans described earlier may be followed for top-working such trees to ensure proper distribution of pollinizer trees.

Temporary Aids of Pollination

Top worked trees and branches will take about two to three years to be able to flower and act as pollinizers. Bouquets and pollen inserts and even hand pollination can be used during the intervening period and whenever there is need for providing additional pollination.

Bouquets

Flowering branches of the pollinizer variety are placed in a water container and hung evenly distributed throughout the orchard. Care should be taken that the bouquets are changed. These should be kept in a fresh condition by changing them and keeping the container full of water.

7

NUTRITION REQUIREMENTS

Orchard productivity and quality of production largely depends upon nutrition made available for maintenance and production. Trees bearing heavy drops need more nutrition. The quantum nutrition also depends upon the type of soil and availability of nutrients in it. Soil analysis and leaf analysis are essential for determining the requirements of the tree. The nutrient removal by important group and crops is given in the chart. However, normal dozes of fertilizer and manures recommended for important crops are also given in the statements. Those are merely guidelines and must be confirmed through leaf and soil tests. Balanced nutrition in horticulture crops is as important as human beings. Each tree should be treated as individual and nutrition determined keeping in view, the age, productivity, soil fertility and deficiencies, if any.

Trials at Chaubattia station have shown that application of fertilizers resulted of gain in productivity by 44 per cent, colour development by 190 per cent, and sweetness by 11 per cent.

The extent of nutrients removed by some important crops given in table below is an indication of the importance of nutrition.

Nutrient Removal by Some Horticulture Crops

S.No.	Crop	Yield T/ha	N	PO	KO	MgO	S
1	2	3	4	5	6	7	8
1.	Apple	25	100	45	180	40	—

(Contd.)

1	2	3	4	5	6	7	8
2.	Banana	40	250	60	1000	140	15
3.	Citrus	30	100	60	350	40	30
4.	Grape	20	170	60	220	50	30
5.	Mango	15	100	25	110	75	—
6.	Papaya	50	90	25	130	15	10
7.	Pineapple	50	185	55	350	110	20
8.	Potato	40	175	80	310	40	20
9.	Cassava	40	150	70	350	40	20
10.	Cabbage	70	370	85	480	60	8
11.	Carrot	30	125	55	200	30	—
12.	Cauliflower	50	250	100	350	30	—
13.	Cucumber	40	70	50	120	60	—
14.	Egg Plant	60	175	40	300	30	10
15.	Okra (*Bhindi*)	20	60	25	90	35	10
16.	Onion and Garlic	35	120	50	160	15	20
17.	Pumpkin	50	90	70	160	40	—
18.	Radish	20	120	60	120	30	—
19.	Spinach	25	125	45	200	35	—
20.	Tomato	50	140	65	190	25	30

Average values taken from various sources summarised by Kemmler, G. and Hobt in *Pottassium a Product of Nature*, Kali-Und-Salz Ag. F.R. Germany.

MANURE, FERTILIZERS RECOMMENDED FOR VARIOUS FRUITS UNDER AVERAGE CONDITIONS

The nutritional requirements of the Trees depend upon its size, growth, productivity, age and the fertility of the soil, etc. We have separately given the amounts of nutrients by some of the important crops. However, in this table we are giving nutritional requirements under average conditions. For heavily bearing trees these can be increased.

Fruit	Age (years)	Farm yard manure kg/plant	Nutrient Basis (g/plant)			Fertiliser basis(g/plant)			Time and method of application
			N	P_2O_5	K_2O	Can	Superphos-phate	Murate of potash	
(1)	(2)	(3)	(4)	(5)	(6)	(7)	(8)	(9)	
Apple	1	10	70	35	70	280	225	120	Apply FYM during December-January along with Phosphates and Potash. Apply half dose of N 2-3 weeks, before flowering and the second half dose one month after the first application.
	2	20	140	70	140	560	450	240	
	3	30	210	105	210	840	675	360	
	4	40	280	140	280	1120	900	480	
	5	50	350	175	350	1400	1125	600	
	6	60	420	210	420	1680	1350	720	
	7	70	490	245	490	1960	1575	840	
	8	80	560	280	560	2240	1760	960	
	9	90	630	315	630	2520	1985	1080	

(Contd.)

(1)	(2)	(3)	(4)	(5)	(6)	(7)	(8)	(9)	
	10	100	700	350	700	2800	2210	1200	In heavy bearing orchards 1 per cent urea should be sprayed immediately after harvest, Pre fall application of Urea (5%) should be given to increase the fruit set.
Pear	1-3	10-20	50-150	32-96	90-270	200-600	200-600	150-450	Apply all FYM, P and K
	4-6	25-35	200-300	128-192	360-540	800-1400	800-1200	600-900	in December. Half of N
	7-9	40-50	300-450	224-298	630-810	1400-1800	1400-1800	1050-1300	can be applied before
	10 and above	50	500	320	900	2000	2000	1500	flowering and other half after first set.
Peach/Plum									
Almond	1	10	50	25	60	200	160	100	Apply FYM during
	2	15	100	50	120	400	320	200	December-January with
	3	20	150	75	180	600	480	300	P&K. Apply half N in
	4	25	200	100	240	800	640	400	spring before flowering
	5	30	250	125	300	1000	800	500	and rest half after
	6	35	300	150	360	1200	960	600	one month under

(Contd.)

(1)	(2)	(3)	(4)	(5)	(6)	(7)	(8)	(9)	
	7	40	350	175	420	1400	1120	700	rainfed condition N
	8	45	400	200	480	1600	1280	800	applied in one lot
	9	50	450	225	540	1800	1440	900	15 days before bud break.
	10-above	55	500	250	600	2000	1600	1000	
Citrus	1	10	80	50	60	320	315	100	P&K should be added if
	2	15	160	100	120	640	635	200	indicated by soil test
	3	20	240	150	180	960	950	300	and are added with
	4	25	320	200	240	1280	1206	400	FYM and applied in
	5	30	460	250	300	1600	1580	500	December end. N in two
	6	35	480	300	360	1920	1900	600	doses first half in
	7	40	560	350	420	2240	2215	700	February, second half in
	8	45	640	400	480	2560	2530	800	April-May. If rainfed N
	9	50	720	450	540	2880	2850	900	be applied on the adwent
	10	55	800	500	600	3200	3165	1000	of monsoon.
Mango	1	10	25	16	60	100	100	100	Apply FYM and P in
	2	20	50	32	120	200	200	200	December and apply 1 kg
	3	30	75	48	180	300	300	300	of CAN and K in February
	4	40	100	64	240	400	400	400	Apply 1 kg can in June.
	5	50	125	80	300	500	500	500	

(Contd.)

(1)	(2)	(3)	(4)	(5)	(6)	(7)	(8)	(9)	
	6	60	150	96	300	600	600	600	
	7	70	175	112	420	700	700	700	
	8	80	200	128	480	800	800	800	
	9	90	225	144	540	900	900	900	
	10-above	100	250	160	600	1000	1000	1000	
Litchi	1	10	60	30	30	240	190	50	Apply FYM, P&K in end of
	2	15	120	60	60	480	380	100	December. Apply half N
	3	20	180	90	90	720	570	150	in mid February and
	4	25	240	120	120	960	700	200	other half in end-April.
	5	30	300	150	150	1200	950	250	
	6	35	360	180	180	1440	1140	300	
	7	40	420	210	210	1680	1330	350	
	8	45	480	240	240	1920	1520	400	
	9	50	540	270	270	2160	1710	450	
	10-above	55	600	300	300	2400	1900	500	
Guava	1 to 9 per year	10	50	40	90	200	250	150	FYM in December Fertilizer, in two doses one
	10-above	50	500	400	900	2000	2500	1500	in May-June, other in Sept.-Oct.
Papaya	Half kg of N P&K (15 x 15 x 15) should be applied twice a year during February and August alongwith 20 kg FYM. No fertilizer should be added in the year of planting.								

NUTRIENT CONTENTS OF DIFFERENT FERTILIZERS

	Fertilizers	Nutrient Content (%)		
1.	Ammonium Chloride	25.0	—	—
2.	Ammonium Sulphate	20.5	—	—
3.	Ammonium Sulphate Nitrate	26.0	—	—
4.	Calcium Ammonium Nitrate	25.0	—	—
5.	Urea	46.0	—	—
6.	Superphosphate (Single)	—	16.0	—
7.	Superphosphate (triple)	—	46.0	—
8.	Ammonium Phosphate	20.0	20.0	—
9.	Diammonium Phosphate	18.0	46.0	—
10.	Diammonium Phosphate	20.0	48.0	—
11.	Ammonium Phosphate Sulphate	16.0	20.0	—
12.	Ammonium Phosphate Sulphate	19.5	19.5	—
13.	Urea Ammonium Phosphate	28.0	28.0	—
14.	Ammonium Phosphate Sulphate Nitrate	20.0	20.0	—
15.	NPK (15:15:7.5)	15	15	7.5
16.	NPK (15:15:15)	15	15	15
17.	NPK (12:32:16)	12	32	16
18.	Dicalcium Phosphate	—	32.0	—
19.	Sulphate of Potash	—	—	48.0
20.	Murate of Potash	—	—	60.0
21.	Bone-meal (raw)	3.4	20.25	—
22.	Bone-meal (steamed)	1.2	25.30	—

CONVERSION OF PURE NUTRIENTS INTO QUANTITIES OF FERTILIZERS

Percentage of nutrient in Fertilizer	Quantity of Fertilizer required rate of the application of the nutrients					
10	20	30	40	50	60	
10	100	200	300	400	500	600
12	85	170	250	330	420	500
15	65	130	200	270	340	400
16	65	130	190	250	310	380
18	55	110	170	220	280	330
20	50	100	150	200	250	300
21	50	100	140	190	240	290
25	40	80	120	160	200	230
26	40	80	120	160	200	230
28	35	70	105	145	180	215
32	30	65	95	125	155	190
46	20	45	65	85	110	130
48	20	40	60	85	105	125
60	15	35	50	65	85	100

NUTRIENT CONTENTS OF MANURES

Manures	Nutrient Content (%)		
	N	P_2O_3	K_2O
Animal refuse			
Dung cattle, fresh	0.3-0.4	0.1-0.2	0.1-0.3
Dung horse, fresh	0.4-0.4	0.3-0.4	0.3-0.4
Dung sheep, fresh	0.5-0.5	0.4-0.6	0.3-1.0
Night-soil, fresh	1.0-1.6	0.8-1.2	0.2-0.6
Poultry manure, fresh	1.0-1.8	1.4-1.8	0.8-0.9
Sewage raw, fresh	2.0-3.0	—	—

Sewage sludge, dry	2.0-3.3	1.0-5.0	0.2-0.5
Sewage sludge, activated, dry	4.0-7.0	2.1-4.2	0.5-0.7
Urine cattle	0.9-1.2	Traces	0.5-1.0
Urine horse	1.2-1.5	Traces	1.3-2.5
Urine human	0.6-1.6	0.1-0.2	0.2-0.3
Farm, factory and habitation wastes			
Compost rural, dry	0.5-1.0	0.4-0.8	0.8-1.2
Compost urban, dry	0.7-2.0	0.9-3.0	1.2-2.0
Farmyard manure, dry	0.4-1.5	4.0-5.0	0.3-1.9
Filter, Press Cake	1.0-1.5	4.0-5.0	1.0-1.7
Plant residue			
Husk, Groundnut Straw and Stalk	1.6-1.8	0.3-0.5	1.1-1.7
Hulls rice	0.3-0.5	0.2-0.3	0.3-2.5
Bajra	0.65	0.75	2.50
Cotton	0.44	0.10	0.66
Jowar	0.40	0.23	2.77
Maize	0.42	1.57	1.65
Paddy	0.36	0.08	0.71
Wheat	0.53	0.10	1.10
Sugarcane trash	0.35	0.10	0.60
Green Manures Fresh			
Dhaincha (*Sesbania aculeata*)	0.62	—	—
Guar, Cluster bean (*Cymposis tet ragonoloba*)	0.34	—	—
Sunhemp (*Crotolaria juncca*)	0.75	0.12	0.61

Source: Handbook of Agriculture (ICAR), year 1990.

8

SOIL AND WATER CONSERVATION IN SITU

Ultimate objective of any land based technology should be that the soil should stay put wherever it is (no erosion). The rain water damage should be minimised (or at naught) during excessive rainfall. The water should be absorbed in soil to its optimum capacity and conserved for use by the plants in dry season. This is what soil and water conservation in Situ actually means. At the same time technology should be universally adaptable, cheap, sustainable, replicable and dynamic, so that it is not only adopted by all but also takes care of the soil and water by adjusting (dynamic) to the future needs of the soil and water. It should be of effective demonstrative value within 2-3 years.

It has been estimated that in India on an average about 14 tonnes of soil is being lost every year from one hectare. Total annual soil losses are estimated to by 5333 million tonnes. The land resources are decreasing while population dependent on land are increasing. More than 30 per cent of the area in India is hilly with varying slopes. While about 31 per cent constitute rolling hills. 21 per cent are steeply dissected mountains. The rolling and steeply dissected hills with high degree of slopes are much more vulnerable to erosion particularly due to heavy rains ranging from 300 to 4000 mm. This area, therefore, needs special treatment if the land resources are to be saved for posterity.

The Technology Components

(a) Planting of Vegetables Hedges on Contour

Depending upon the gradient or slopes of land, vegetative hedges of appropriate species are to be planted at suitable intervals. The hedges become dense in 3-4 years and are able to sieve soil out of runoff water. The sieved soil accumulates on the upper side of the hedge and if distance between two hedges is not wide, a natural terrace will be formed in due course of time. This natural levelling between two hedges is better than done by man by mechanical means. It will be permanent vegetative dynamic structure needing no repairs or upkeep. Terrace making in such slopes with human labour and machines is costly, and not recommended.

Since the hedges are on contour the water flow which carries the soil will (in a few years) spread throughout the contour line or hedge instead of passing through small waterways made by water itself because of uneven land surface. This process or action of water ultimately makes gullies and streams within the fields. The spread of water throughout the breadth reduces its speed or velocity as well as capacity for eroding and carrying the soil. By the time the flow gathers speed, it is checked by another hedge. Thus soil is retained and stabilised.

Equal spread and slow movement of water allows more time to the soil for absorbing water to its capacity which otherwise does not happen and results in flash floods or quick movement of water flow if not checked by such barriers.

As a result of these measures, water conservation (or retention) in the soil is more and plant nutrient losses are reduced. At the same time lower areas are also benefited with improved ground water level and reduction in floods. Illustrations of the important components of the technologies are given in charts 1 to 4. These are self explanatory.

(b) Basic Requirements of Planting Material

A good and effective vegetative hedge plant should have the

following characteristics:

(i) Should be hardy and form a dense fence at ground level so that the soil is sieved out of the runoff water.
(ii) Should not be an alternate host for pests and diseases of the main crop.
(iii) Should not compete with the main crop for moisture during critical periods of crop growth.
(iv) The weeds should be non-violable so that it does not spread as a weed in the area.
(v) Should not be a fire hazard.

Species including Khus, Agave (*siliana*) and lemon grass can be tried before large scale application.

(c) Planting of Trees on Contour

The trees of horticulture species should be planted on contour and not along the slope as is done now. This (present method) is totally wrong method and besides causing erosion and runoff of most of the water, is also difficult to manage. The spacing between two trees in a line on the contour may be less than between two contour lines. It will almost be rectangular plantation, not a square one as is generally the case now. The density of plants should be optimum.

(d) Distance of Planting Vegetative Hedges

The spacing between the vegetative hedges largely depends upon the slope or gradient. More the slope less the distance between two lines. If the slope is 5 per cent (or say drop of 5 metre in 100 metre length of the slope) the hedges will be 60 metre apart horizontally. On a 2 per cent slope the hedges would be 100 metre apart. However, if the slope is 57 per cent the distance would be 7 metres.

CHART - I

MOISTURE CONSERVATION IN SITU TECHNOLOGY

CORRECT CONTOURS

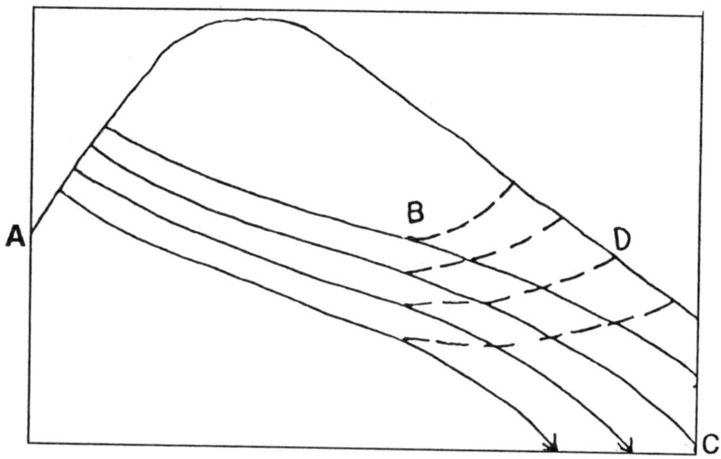

WRONG CONTOURS

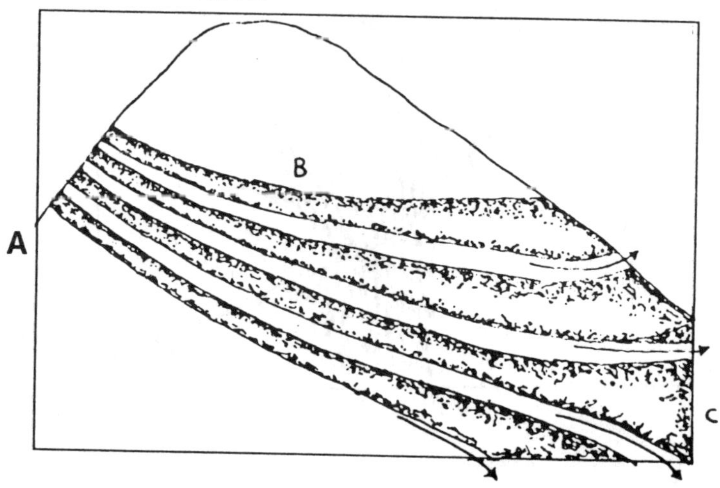

VEGETATIVE SYSTEM OF SOIL MOISTURE CONSERVATION IN SITU

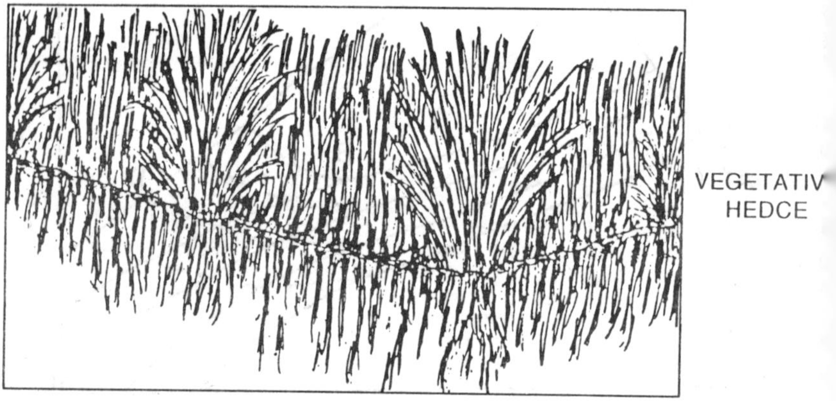

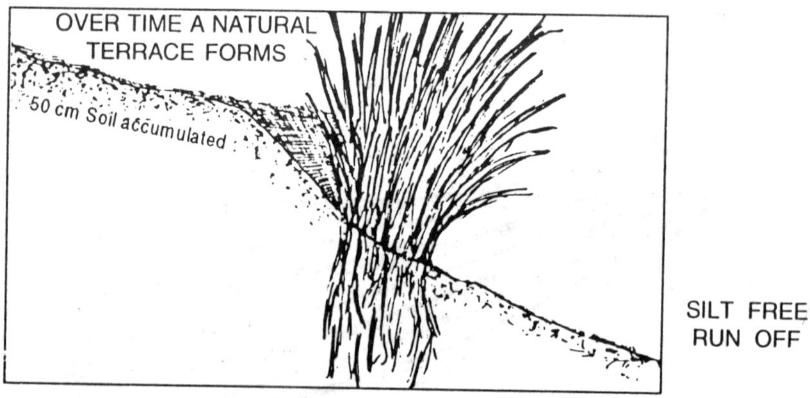

SOIL AND MOISTURE CONSERVATION IN SITU FOR HILL SLOPE FARMING

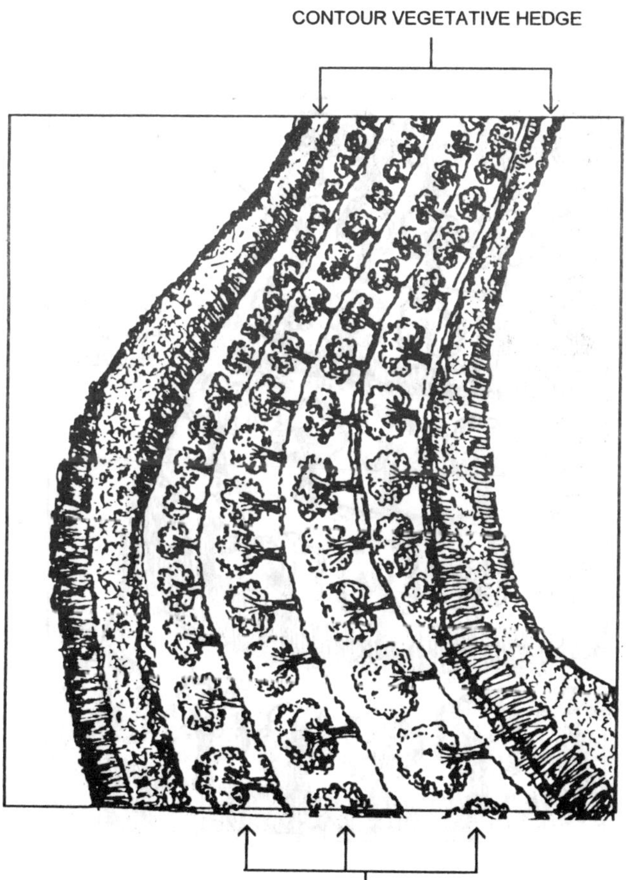

SOIL AND MOISTURE CONSERVATION IN SITU FOR HILL SLOPE FARMING

ANYWAY

People are unreasonable, illogical
and self centered. **Love them anyway.**
If you do good, people will accuse you
of selfish ulterior motives. **Do good anyway.**
If you are successful you win false
friends and true enemies. **Succeed anyway.**
The good you do today will be
fogotten tomorrow. **Do good anyway.**
Honesty and frankness make you
vulnerable. **Be honest and frank anyway.**
People favour underdogs, but follow only
top dogs. **Fight for some underdogs anyway.**
What you spend years building may
be destroyed overnight. **Build anyway.**
People really need help but may attack
you if you help them. **Help people anyway.**
Give the world the best you have
and you'll get kicked in the teeth.
Give your best anyway.

—Anonymous

9

ORGANIC FARMING/HORTICULTURE

Environment and ecology related issues have been exercising the mind of people the world over. Use of chemical fertilizers, pesticides and various other chemicals have depleted soil health and created environmental hazards to the extent that corrective measures have become essential. In the activity relating to farming and horticulture emphasis on organic farming is being increasingly stressed.

What is Organic Farming

Organic farming has been differently defined by various people. Most accepted definition presented by the United States Department of Agriculture (USDA) is that "Organic farming is a production system which avoids or largely excludes the use of synthetically compounded fertilisers, pesticides, growth regulators and livestock feed additives. To the maximum extent feasible organic farming system rely upon crop rotations, crop residues, animal manures, legumes, green manures, off farm organic wastes, mechanical cultivation, mineral bearing rocks, and aspects of biological pest control to maintain soil production and tilth, to supply plant nutrients and to control insects, weeds and other pests".

The major aim of the organic cultivation is to secure the advantages and economics of working with nature by bringing in more and more plant foods into circulation and to obtain greatest yields of most nutritious crops while maintaining fertility of the soil. Composing, Vermicomposting, Biomass and recycling of weeds and agriculture biomass are integral parts of organic farming particularly under intensive methods of cultivation.

Soil Plant Relationship

It is essential to understand soil plant relationship particularly by an organic farmer. Soil provides reserves of water and nutrients and provides the matrix for biological processes involved in nutrient cycling. In return plants play a role in soil formation, provide energy and sustain biological process. Even up to 80 per cent photosynthates are translocated to roots. High proportion of these is used to feed organism in rootzones and the soil. A large range of organisms are part of life of the soil. A 'thimble full' of soil may contain billions of bacteria and actinomycete and several kilometres of Fungal lineaments. These rich assortment of organisms benefit tree growth immensely through decomposition, capture and uptake of nutrients, nitrogen fixation, protection against moisture stress and maintaining soil structure. The variety of soil life range between the smallest living organisms to small mammals.

Over 300 years ago John Baptise Von Helmut planted a willow sapling and watered it for 5 years. The willow gained 36.17 kg weight but only 56.78 grams of soil was lost. The increase in weight evidently came from the atmosphere with the help of the soil organisms.

Green manuring is one of the important practices for improving soil fertility. Wide range of crops are now available for green manuring in orchards. These are also known as fertility promoting crops. The ground is kept covered thereby protecting soil against heavy rain pounding and drying out by strong sunlight. Excess water during the rainy season is evaporated through the process of transpiration. These also help to bring minerals into circulation from store. If the crop is deep rooted it brings up minerals from lower depths. There is substantial weed control and leaching of nutrients is controlled or prevented. The nitrogen content increases if the crop is leguminous. It acts as a mulch if left spread on the ground with all its attendent benefits which include prevention of nitrogen loss on account of oxidation and encouragement of surface living organisms (such as nitrogen fixing). Green manuring of suitable crops should, therefore, form an integral part of organic farming system. Yield of oranges increased 425, 387 and 383 per cent with green manuring of different crops. For instance alfa-alfa may add 217 kg. of N, Cowpea 101 kg, Bereem 140 kg. and Sesbenia 30 kg. N, 8 kg. P_2O_5 and 68 kg. K_2O.

There is close co-relation between the health of the soil and health of the human beings and animals. Healthy soil is full of life and produce healthy crops, fruits and vegetables. Billions of micro-organism exist in a handful of healthy soil. Health of the soil is directly related with their activity. For this, conditions must be congenial which can only be provided by organic farming. Most of the chemicals have adverse effect on soil life.

Use of chemicals (fertilisers and pesticides) has been found to be harmful. In many high fertiliser consuming areas ground water has been found to be polluted. Nitrate contents of the drinking water have increased. The marginal productivity of N and P has atleast dropped to 1/3rd in last 2 decades. Similarly extensive use of pesticides has resulted in contamination of all kinds of foods (fruits, vegetables, cereals and even milk).

Transport of fertilisers to the hills (in the interior) is very costly. Moreover, their effectiveness on slopy soils is not so efficient. Organic farming is, therefore, the best solution not only today but also for posterity. Most of the orchards are planted on slopes. The area in between is not recommended to be earthed up as it causes serious erosion problems. The (sod) grass-legume mixtures in the interspaces is the best. Local grasses should be encouraged to grow by removing useless and unpalatable bushes. This can be mixed with legumes. This will depend upon the height. Red, pink and white clovers are the best for temperate region. Other species like alfa-alfa and kudzu are suitable for all zones. Similarly, thick clover, sweet clover, velvet beans, etc. are suitable for mild hill and low elevation areas. Alfa-alfa can be grown in all zones.

Mulching

In addition to sods in the interspaces, mulch should be used for moisture conservation (unirrigated areas) prevention of weeds and soil erosion from the tree base. This is another kind of organic farming as the mulch ultimately is turned into manure. Excess sod can be used as mulch and even composting.

Composting

In addition to regeneration and recycling of the plant growth within the orchard without removal from the site (except mulching) the nutrients can be supplied to the trees through scientifically prepared compost. Again best compost is obtained with the abundant population of earthworms. This is called vermicompost. The earthworm's activity depends upon the media. These are not active (though present) in saline media or chemicals. Each farmer has excess Biomass which be used as litter for animals or even excessive growth of grasses, shrubs and even trees. All this material along with dung should be used for composting. In USA and now in India several firms are raising and marketing earthworms.

Several models of compost pits are available including that of using bricks/stones. However, one can get good results by having a pit of the size of 2 x 3 metres or so. Ordinarily height is kept as one metre or so. All types of animal yard litter (cowdung and Biomass) is put in this pit. Other plant wastes are also added. After every 1/3 metre and filling, cowdung solution (cowdung + water) is spread over it and preferably topped with dry soil. This is again repeated twice. Moisture level is generally kept 15 to 20 per cent. Because of the low temperature in higher hills-it may take 6-8 months for the formation of well rotted compost. In orchards the compost is applied only in winter. As such even one year cycle will suffice. For vegetables one can use after 6-8 months, if it is ready.

In some recommendations, a good deal of soil is suggested to be added to compost for covering of each layer. This can be avoided as scraping of soil causes erosion in hills and compost cycle can be afforded to be yearly.

Vermiculture now available with some organisations can be added to improve the density of the population of earthworms.

10

TRAINING AND PRUNING

The main objectives of training and pruning are:

(i) Have a proper shape and form of the tree.
(ii) To increase trees and its limits strength capable of bearing heavy crop loads without damage of breakage.
(iii) Regulate bearing habit of the tree.
(iv) To improve health, vigour, yield and quality of fruit.
(v) To make orchard operation like pruning, pickings, control of pests and diseases easier.

The training and pruning is necessary in case of all trees. Some need more and others less pruning. It should, however, cause minimum interference with the natural growth habit of tree and its fruiting.

The training and pruning itself is a very vast and complex and several books can be written. The purpose of this guide will be served if a short description with illustrations is presented. Training is generally done to develop a strong and well balanced frame work of the tree so that it can bear optimum load of crop without breaking. This is called pruning for form and is done during the first few formative years of the tree. Pruning in subsequent years is done keeping in view the growth and bearing habit of the tree.

For the purpose of this guide, training and pruning of apples and peaches which are two typical examples of heavy fruit on old wood (spurs) and last seasons growth (peach) are presented as broad guidelines.

Important Definitions and Terms

Some readers may not be well versed with the terms commonly used. Knowledge of these terms is necessary pre-requisite for properly understanding the following text. These are, therefore, being briefly defined and illustrated in most of the cases in Figs. 1 and 2.

1. **Collar** —The point where root-stock (roots) and the scion (trunk) join.
2. **Trunk** — This part of the tree extends from the collar to the place where first branch emerges.
3. **Leader** — Extension of the trunk above the branches.
4. **Scaffold or main limbs** — Branches which originate directly from the scaffold limbs.
5. **Secondary branches or laterals** — Branches which originate directly from the scaffold limbs.
6. **Stems** — The wood or branches beyond secondary branches.
7. **Twigs** — The terminal stems from the preceding season's growth.
8. **Thinning out** — The removal of a whole twig or small branches from the place of origin.
9. **Heading back** — Cutting out a portion of a branch of a twig.
10. **Heavy pruning** — This consists of pruning of equal to or more than ¼th of annual shoot growth. Light pruning is when it is less than ¼th of the annual twig growth.
11. **Spur** — It is a very small fruit bearing branch. It makes very short annual growth. Each annual growth is represented by a ring left by leaf scars.
12. **Water sprouts** — These are the fast growing succulent shoots arising from the crown, trunk or main branches or even from the laterals. Because of the quick vegetative growth these come into bearing very late.
13. **Framework** — The main above ground branching system of the tree consisting of the leader, scaffold branches and

the laterals.

14. **Whip** — One year old unbranched nursery plant ready for planting in the orchard.
15. **Tertiary buds** — Inactive buds on the trunk or any other limb of the tree. When the tree gets a severe set back and normal and dormant buds are damaged or exhausted. These incipient buds, also commonly known as tertiary buds, become active and sprout.
16. **Height of the Head** — The distance or the clearance between the ground and the main limb branching from the Trunk.
17. **High Headed Trees** — The trees where the scaffold limb branch out 137 cms, or so (4½ feet) or above from the ground are known as high headed trees.
18. **Low Headed Trees** — The trees in which case the scaffold limb branch out 75 to 90 cms, (2½ or 3 feet) from the ground are known as low headed trees.

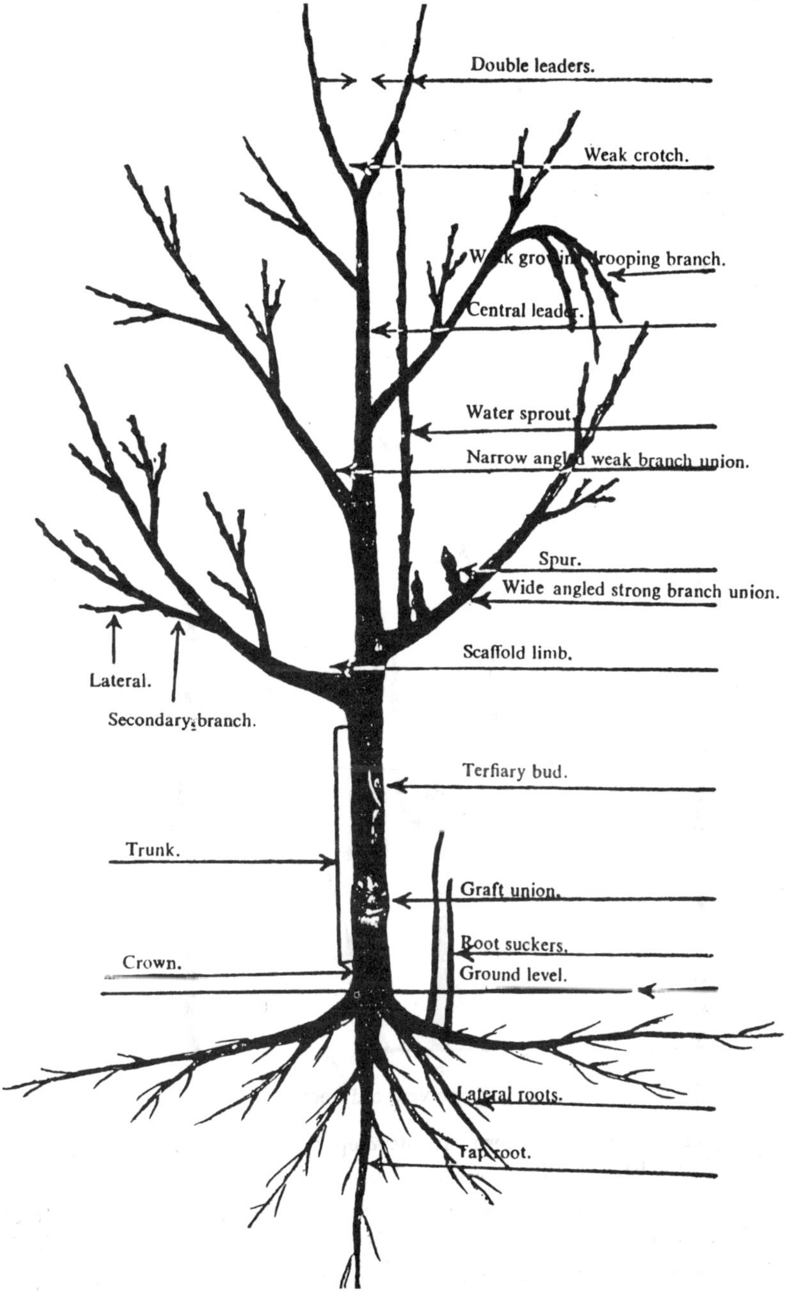

FIG. 1: A Diagram of the Tree Showing Various Parts

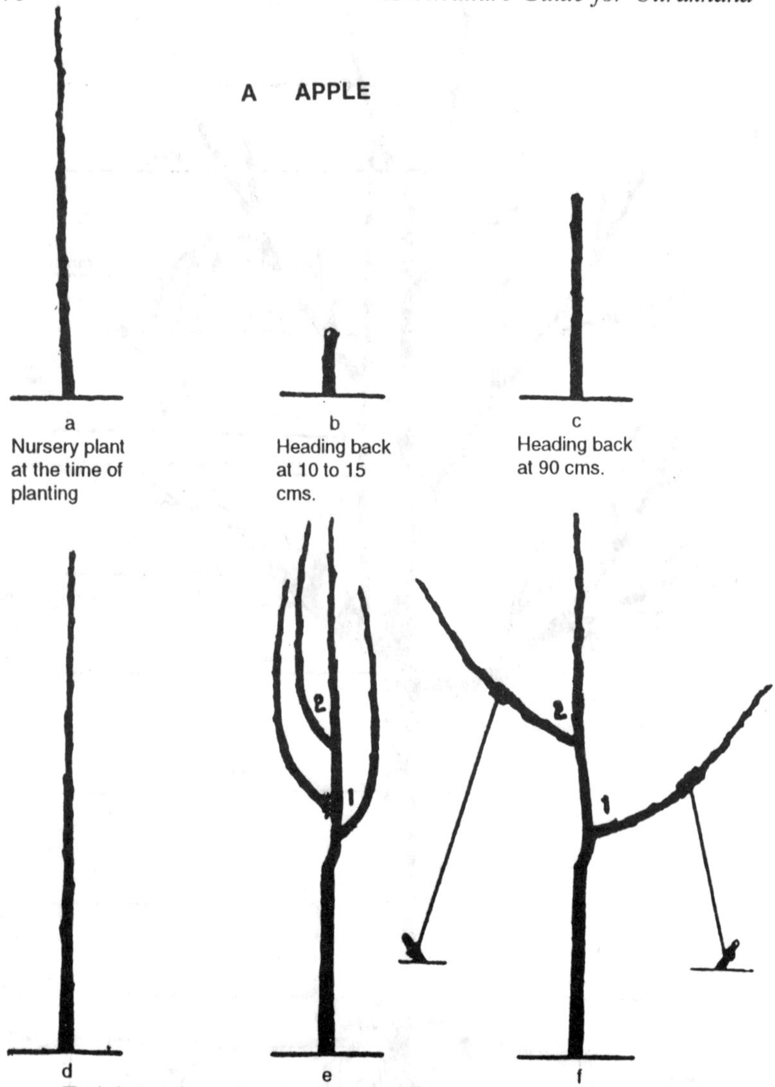

Training and pruning of one year old tree.

(d) One year old tree in dormancy developed from the plant at b which was headed back at 10 cms. height last year.

(e) The tree at C has produced three branches in addition to the leader before pruning.

(f) The tree at e after dormant (winter pruning). The branch marked in e has been removed as it was too near the to scaffold limb No. 1. Note the direction of the branches which wer tied.

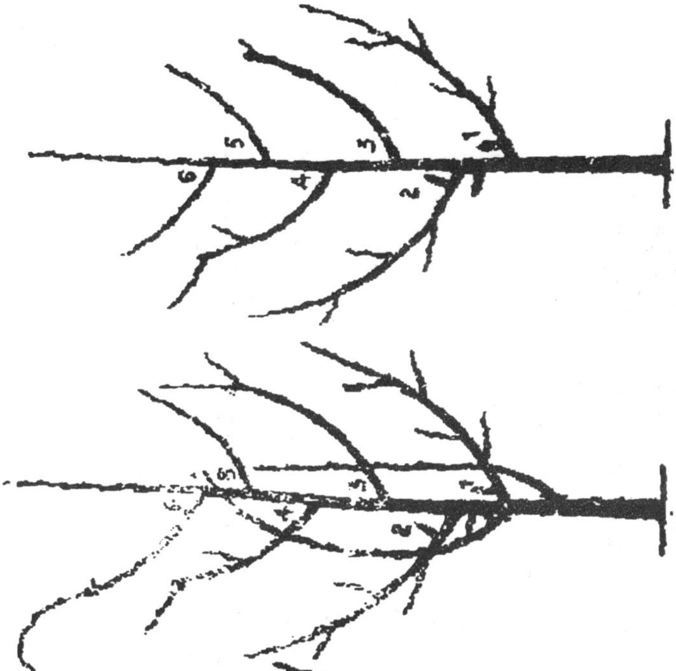

3. Two year old tree in dormancy

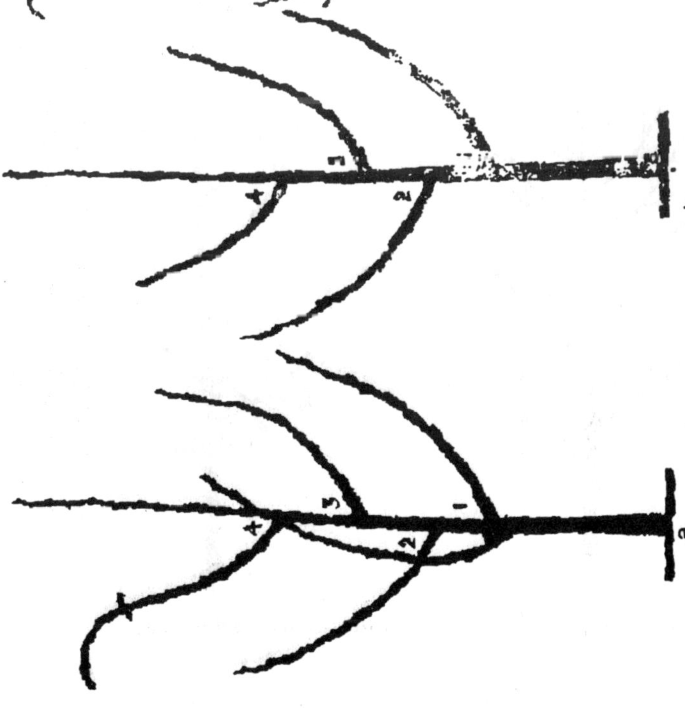

5. Three year old tree.

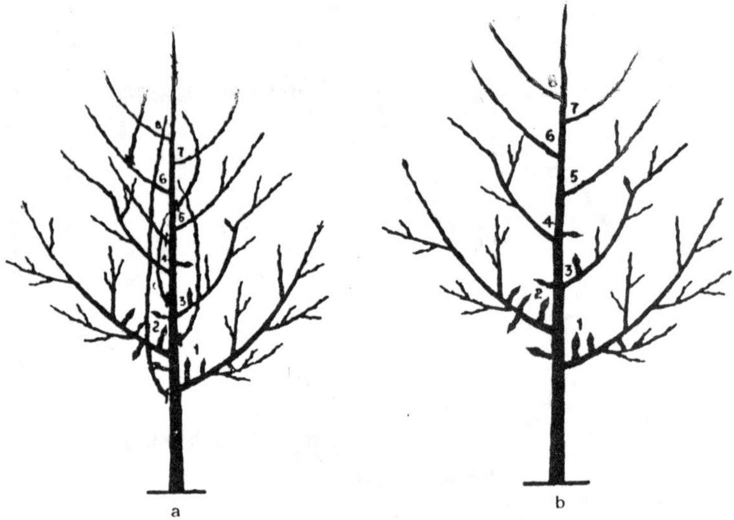

7. Four year tree before (a) and after (b) dormant pruning

9. Five year tree before (a) and after (b) dormant pruning

23. Modification of the leader

Central leader has been removed near a strong side branch No. 10. The open space in the top of the tree will be filled up in couple of years with the growth of shoots from top branches. The expected shoots have been drawn in dotted lines.

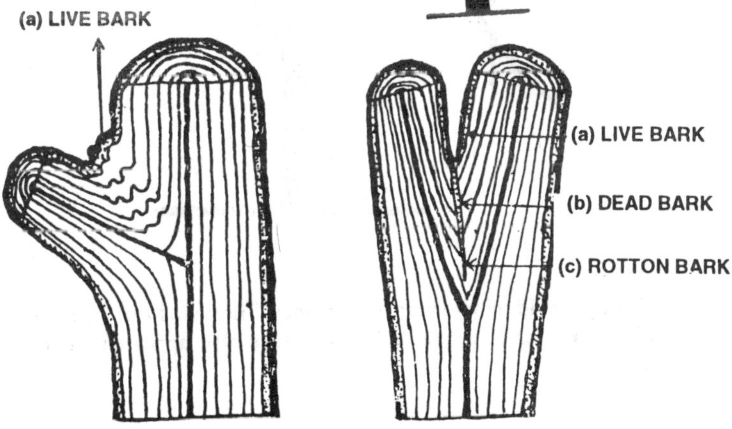

Left : A wide angled branch having a strong union with the mother limb.

Right : A narrow angled weak crotch. The dead and rotten bark inclusions as shown at b and c weaken the branch which is broken with comparatively much less weight and pressure.

14. Three year old spindle bush in dormancy. There are in all 7 branches well distributed on the spindle. Shoots have also appeared on the main branches.

15. Four year old spindle bush. The tree has almost completed its branching system and is ready for bearing a good crop load. The branches have been well distributed on the central leader. Further growth will mainly consists of the extension of the branches on the periphery of the main branches and additional shoots on the existing limbs.

16. Six month old Golden Delicious tree being trained as spindle bush.

17. First summer growth of a spindle bush. The two main branches on the central leader (spindle) have been tied so as to force these into horizontal position. The shoots in dotted lines are removed as soon as possible in early summer. These are not required, because the lower shoots are too near the ground and the upper one is directed towards the central leader.

18. One year old Spindle bush tree in dormancy. Because of the judicious handling during the preceeding summer the tree does not require any pruning except for the tipping of the leader, which had become weak at the top, so as to make it strong.

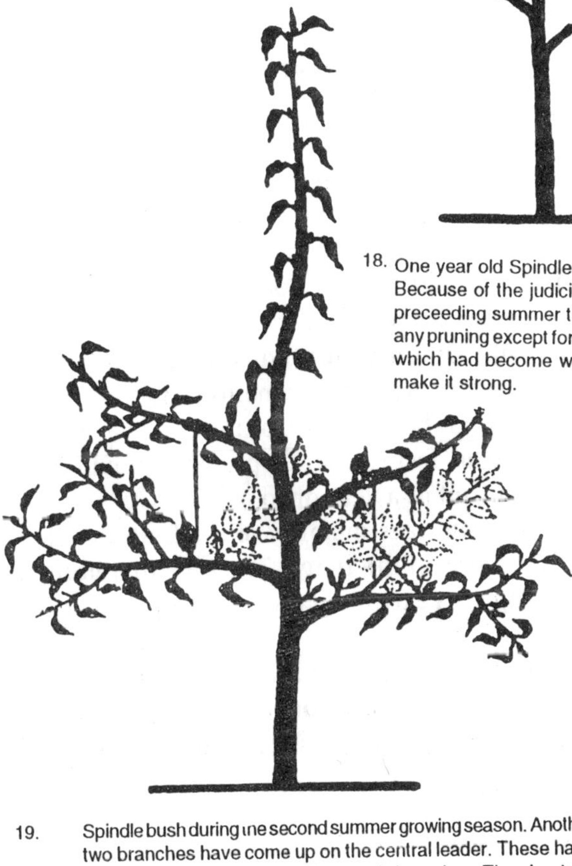

19. Spindle bush during the second summer growing season. Another two branches have come up on the central leader. These have been tied as in case of the first year's branches. The shoots in dotted lines are either crossing each other or otherwise misplaced. These are to be removed as early as possible.

B. Peach

Pruning is a very controversial subject of orchard tree management. it differs from the training and pruning of Apples and Pears. It must, therefore, be understood thoroughly before undertaking the operation.

Peach trees differ from apple in growth habit, fruit bearing habit and age of bearing. The cultivation of the peach, therefore, differs from that of apple. This difference is the greatest in respect of pruning. The bearing habit of peach is almost the opposite of apple. Whereas most of the apple varieties bear on old wood (spurs) the peach bears on the last season's growth. Spurs are produced in some cases and varieties but these are different from apple spurs in that they are straight and bear from laterals instead of terminal buds. The terminal buds are always vegetative. In case of the apple the interest of the farmer lies in retaining the spurs and old wood for some years. This is not so in case of peach. Here the part of the shoot and branch having once borne fruit becomes barren. The fruit is always borne on the last season's growth. Thus the fruit and fruiting wood for the next years crop are developing at the same time and the same season. Each peach twig has both the fruit bearing and vegetative functions. A grower who forgets peach pruning should also forget of getting peach fruit in a year or so.

On most of the young trees majority of the fruit buds are developed towards the tips of the twigs. There is, therefore, a tendency of the bearing area extending towards the periphery of the tree every year.

Thus the bearing area gets further away from the centre of the tree every year in case the tree is not pruned properly. Such trees have more limb breakage and less crop per unit spread of the tree.

Pruning of Mature Trees

Peach trees become almost fully grown in about five years. The

Fig. 1. A. Peach tree at the time of planting
B. Heading back at 60 to 75 cms. height after planting

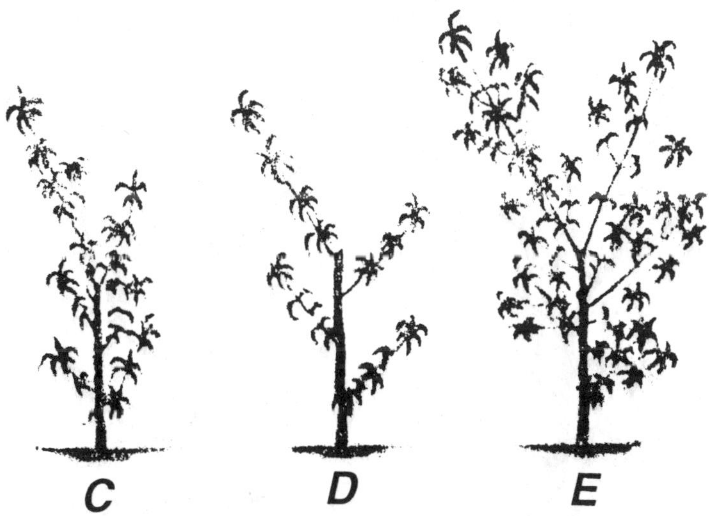

Fig. 2. C. Growth during first two months
D. First summer pruning after 2 months of summer growth.
Note: Only four well spaced branches have been retained.
E. The same tree after full summer growth and before leaf fall (one year old tree)

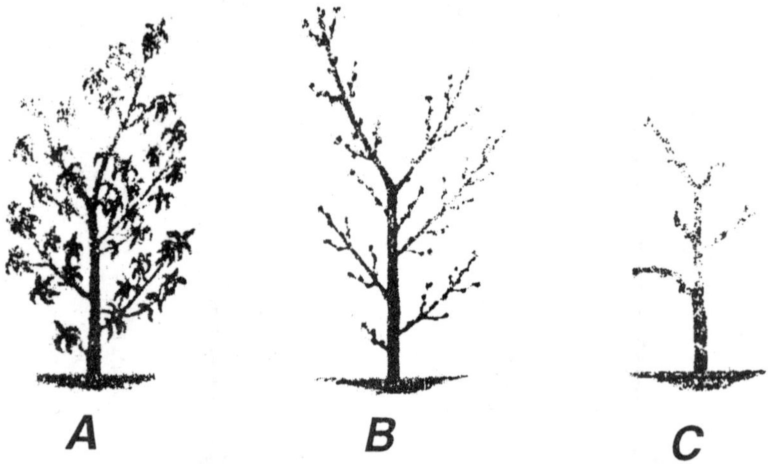

Fig. 3 A. One year old tree before leaf fall
B. The tree at A in dormancy, before pruning
C. The tree at B after winter pruning. Note only four branches have been kept and these have been headed back at least by 50 percent. These will form the main framework of the tree.

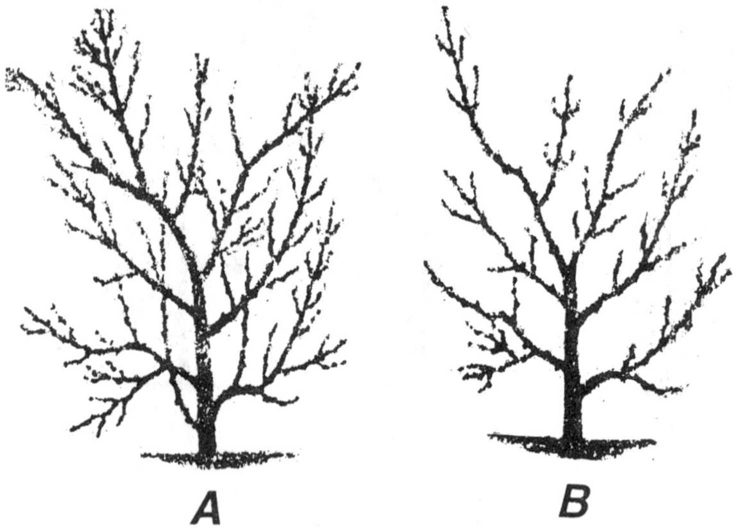

Fig. 4 A two year old tree before (A) and after (B) pruning. Water sprouts mis-placed shoots arising in between the main frame work, shoots crossing others, etc., have been removed during the second winter pruning. The branches have also been headed back.

trees will be continued to be pruned as explained under Figs. 5A and B. Care has, however, to be taken that the branches are judiciously headed back so that there is a balance between the current season's fruit crop and the growth which will be bearing fruit next year.

The following points should be taken into consideration while pruning a mature peach tree:

1. First of all, remove all dried and diseased twigs and wood.
2. The twigs which have borne fruit in the preceding season and have no or very limited extension growth should also be removed.
3. Then remove crossing and shading branches which are interfering with the growth of more useful branches or twigs.
4. The remaining branches and shoots should be headed back by normally one third of their growth. So as to keep the tree compact, allow more vegetative growth for the next year's crop and thin the fruit crop.
5. The weak and leggy branches should be more severely headed back. This may be half or even two third of the growth.

While pruning a peach tree the pruner must have an eye (foresight) on the immediate crop and the next year's crop. In case the fruit bud formation is abundant as is usual with peach, pruning and heading back may be more severe than if the fruiting wood is light or scanty. Pruning is also a mode of thinning of fruits in heavily bearing peach trees.

Pruning of Old Trees

After a few year's growth the fruiting area is generally confined to the periphery of the tree particularly in case of the neglected trees and interior of such trees comprises of mostly thick and barren limbs (Fig. 6-A). This can be rectified by severely heading

Fig. 5. A three year old tree before (A) and after (B) pruning. In addition to water sprouts, un-wanted and congesting branches have also been thinned out. The remaining branches have been headed back by about 1/3rd of the last summer growth. The tree had borne some crop during the preceding summer season.

Fig. 6. Pruning of mature and fully grown tree. For the sake of clarity, only one limb has been exhibited. The tree has already borne for some years and the fruiting wood is going away from the centre of the tree. The branches in the limb (A) have been alternatively lightly tipped and severely headed back as shown in *B*. All the limbs should be pruned in the same manner.

Fig. 7 A. The limb at 6 (B) after one year's growth. This has again to be pruned in same manner i.e. alternate branches lightly and severely, with the difference that branches which received heavy pruning last year will get light pruning and vice-versa. This alternate pruning will enable the orchardists to keep his trees compact and get more and better quality crop, throughout the tree spread.

B. The limb at *A* after pruning in winter.

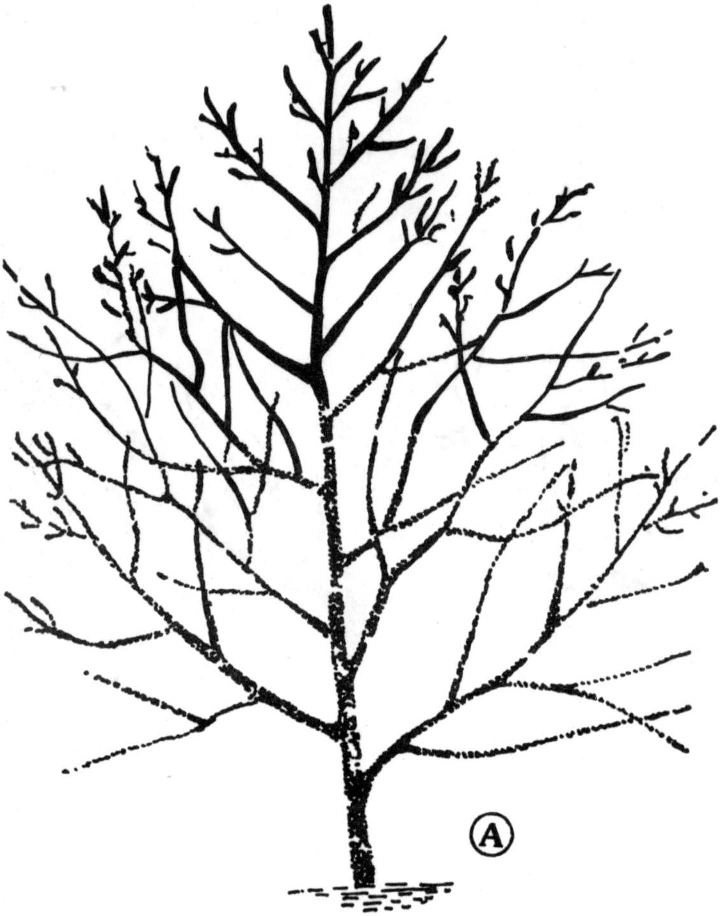

Fig. 8 A. An old tree whose bearing area is confined to the outer periphery and tips of the branches. These branches are liable to breakage and the trees bear less crop per unit area. Such trees should be dehorned as show in Fig 'B'.

back alternative branches or each limb every year.

Deheading or Dehorning

Sometimes well grown trees are severally affected by frost or some good old trees are required to be rejuvenated (Fig. 8A). Such trees are pruned severally. Almost all the tops are headed back to secondary branches. This has been illustrated in (Fig. 8-B). Growth in the following summer may be abundant (Fig. 8-C). Surplus shoots should be thinned out and others headed back. Because of the chances of the death of the dehorned trees and loss of two years crop, deheading should be resorted to when it is a must. Replacement may be better than rejuvenation in majority of the cases.

Fig. 8. B. The tree at 8A after severe heading back. The main limbs have been cut to two to four laterals and each lateral has two to four buds.

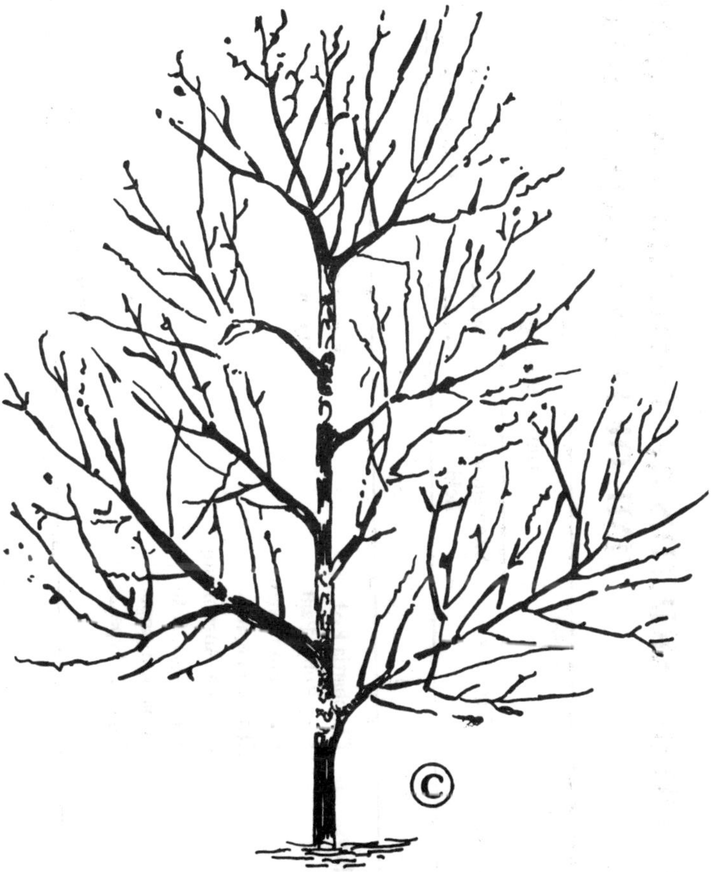

Fig. 8 C The tree at *B* after completing full summer growth.

11

VEGETABLES, THEIR VARIETIES RECOMMENDED WITH DETAILS OF TIME OF PLANTATION AND SEED RATES

S. No.	Vegetable	Region	Varieties	Time of Sowing		Time of Transplantation	Seed Rate kg/ha.
				Straignht in the field	In the Nursery		
1	2	3	4	5	6	7	8
1.	Tomato	Valleys	Chaubattia Red, Pusa Ruby, Pant Bahar, Pusa Early Dwarf, Punjab Chuaara, Angoor latu Arka Vikas, Arka Vishal Kalyanpur Type-1, Type-2	—	February to March	March to April	400-500
		Hills	-do-	—	March to April	April to May	-do-

(Contd.)

1	2	3	4	5	6	7	8
2.	Capsicum	Valleys	California Wonder Bullnose, yellow vilonder, Chinese Giant, Arka Basant Arka Mohini and Elephant Trunk	—	Mid-February to Mid-March	March to April	1.00 to 1.5 kg/ha Arka Gauran
		Hills	-do-	—	March to April	April to May	-do-
3.	Brinjal	Valley and Hills	Pusa Purple long Pusa Purple cluster Pant Samrat, Asad Kranti, Asad B-1 Arka Navneet, Arka Sircish	—	March-April	April to May	500-600 gm/ha
4.	Chilli	Valley & Hills	Pant C-1, Pusa Jwala Chanchal, Local Strain	—	March-April	April-May	1.00-1.200 kg/h
5.	Cauliflower	Mid & high	Snow ball-16, PSB-1 Pant shubbra, Deepali Pusa Synthetic, Jawahar Moti, Early, Kawari	—	Early Var-June to July Late Var-Sept. -Oct.-Nov.	July to August (Early) Oct.-Nov.-Dec.	600-700 gm/ha 400-500 g/ha (Late)

(Contd.)

6.	Cabbage	Valley and Hills	Golden Aere, Pride of India, Early drum head, Chaubattia Early, ARU-Glony and Green Express	—	Early May-June Mid-August-Sept. Late Oct.-Nov.	June-July Sept.-Oct. Nov.-Dec.	600 gm/ha (Early) 500 gm/ha (Mid) 500 mg/ha (Late)
7.	Pea	Valley & Hills	Arket, Bonville, Azad P-1, New Line Perfection	August-Sept. Oct. to Mid Nov.	—	—	100-120 kg/ha (Early) 80-90 kg/ha (Late)
8.	French Bean	Valley and Hills	Chaubattia Top Crop V.L.-1, contender, Pant Anupam, Arka Komal Black Queen, Black Princess	Feb. to June (according to Altitude)	—	—	80-90 kg/ha
9.	Radish	Mid and High Hills	Japanese white, French Break fast, Arka Nishant Crimson Giant, Pusa Chetaki Pusa Reshami	June to Sept.	—	—	10-12 kg/ha.
10.	Carrot	Mid and High Hills	Nantes, Early Nantes, Chantery Long, Pusa Kesar	May to Aug.	—	—	4-4.5 kg/ha

(Contd.)

1	2	3	4	5	6	7	8
11.	Summer Squash	Mid and High Hills	Australian Green Chaubattia long Green Pusa Alankar, Improved green bush Casesta, Cocozells.	Feb. to April	—	—	6-8 kg/ha
12.	Cucumber	Valley and Hills	Japanese Long, Saharanpur Long Green, Poinsettee, Pusa Sanyog, Local Strain, Balam kheera	March to May	—	—	2.5 kg/ha

12

PLANT MATERIAL FOR LANDSCAPING

I. Trees

S.No.	Botanical Name	Common/English Name
I.	**Trees for Peripherial Planting**	
1.	Casuarina equisetifolia	Casureen Beef wood
2.	Eucalyptus citriodora	Sufaida
3.	Grevillea robusta	Silver oak
4.	Polyalthia longifolia	Ashoka
5.	Populus hybrid	Hybrid poplar
6.	Populus nigraitalica	Kashmir Poplar
II.	**Trees for Shade**	
1.	Alstonia scholaris	Scholar tree
2.	Cassia siamea	Kassod
3.	Chakarasia tabularis	Chakarassia
4.	Cedrela tuna	Tun
5.	Ficus lacor	Pilkan
6.	Madhuca indica	Mahua
7.	Mimusops elengi	Mqusari
8.	Pongamia pinnata	Sukhchain
9.	Putranjiva roxburghii	Jivapota
10.	Pttrospermum acerifolium	Kanak Champa
11.	Sweotenia Mahagani	Mahagony
12.	Sachieichera elcosa	Kusum
13.	Thespesia populnea	Tulip tree

III. Flowering trees for wide roads and large compounds

1.	Cassia fistula	Amaltas
2.	Cassia javanica	Javadirani
3.	Cassia nodosa	Pink Modar
4.	Cassia spectablis	
5.	Milletia ovelifoia	Rose wood
6.	Dclonix regia	Gulmohar
7.	Peltophorum peterocarpum	Yellow Gulmohar

IV. Flowering trees for narrow roads and small compounds

1.	Acacia auriculiformis	Australian wattle
2.	Bauhina blakeana	Kachnar
3.	Bauhina triandra	Kachnar
4.	Bauhinia variegata	Maxican silk
5.	Chorisia speciosa	Indian coral tree
6.	Erythrina variegata	Neeli Gulmohar
7.	Jacaranda acutifolia	Queens flower
8.	Lagerstoemia speciosa	Safaid Champa
9.	Plumeria alba	Tulp tree
10.	Thespesia populnea	
11.	Tecomela undulata	
12.	Tebublia rosea	Laharoo

V. Drought resistant trees

1.	Albizzia procera	Siris
2.	Butea monsopperma	Dhak
3.	Cassia fistula	Amaltas
4.	Casuarina equisetifolia	Casureena/Beef wood
5.	Createva nurvala	Barna
6.	Maduca indica	Mahua
7.	Pongamia pinnata	Karanj
8.	Parkinsonia aculeata	Jerusalm thom
9.	Tamarindus indica	Imli

S.No.	Botanical Name	Common Name	Size	Flower colour	Flowering season
1.	Barleria cristata	Barleria	Dwarf	Blue, white orange	Winter
2.	Bougainvillea	Bougainvilea	Dwarftall and purple	White, orange red	March-April and Oct.-Nov.
3.	Caesalpina pulcherrima	Gulmohari	Tall	Orange or Yellow	Summer
4.	Cassia glauca	Cassia glauca	Tall	Yellow	Early winter
5.	Cayopteris incana	Sambalu	Tall	Blue, pink, white	Winter
6.	Cestrum nocturnum	Rat ki Rani	Medium	Green white	All the year round
7.	Daedalacanthus	Nurvosus	Dwarf	Blue	Spring
8.	Euphorbia pulcherima	Poinsettia	Medium Yellow	Red, pink, or yellow	Winter
9.	Ervatamiacoronaria	Double chandni	Tall	White	through out the year
10.	E. divaricata	Single Chandni	Tall	White	-do-
11.	Galphima gracilis	Galphimia	Dwarf	Yellow	-do-
12.	Gardenia jasminoides	Gardenia	Medium	White	Summer
13.	Halelia patens	Hamelia	Tall	Orange, Red	All the year round

(Contd.)

S.No.	Botanical Name	Common Name	Size	Flower colour	Flowering season
14.	Hibiscus rosasinensis	Shoe flower	Tall	Red	Summer
15.	Ixora coccinea	Ixora	Medium	White	Winter
16.	Jasminum multiflorum	Nala	Medium	White	Winter
17.	Jatropha hastata	Jatropha	Medium	Scarlet, pink	All the year round
18.	Lagerstroemia indica	Sawni	Medium to Tall	Pink, white or Mauve	Summer rainy season
19.	Lantana Camara	Lantana	Medium	White, yellow	All the year round
20.	Malvavisuls arboreum	Achania	Medium	Scarlet, red or pink	Winter
21.	Nerium indicum	Oleander	Tall	White, pink or red	Summer
22.	Russelia juncea	Weeping mary	Dwarf	Red	All the year intermitently
23.	Russelia rotundifolia	—	Dwarf	Red	-do-
24.	Swpirea betulifolia	—	Dwarf	White	Spring
25.	Tecoma stans	Tecoma	Tall	Yellow	Throughout the year intermitently
26.	Thevetia peruvina	Peeli kaner	Tall	Yellow	-do-
27.	Jasminum Sambac	Motia	Dwarf	White	Summer

II. Shrubs

It is the careful selection and skitful planting of shrubs that make the environment of place colourful and fill the gap between tall trees and green lawns with bright flowers and green foliage.

III. Climbers

Climbers when trained against the high building walls break the monotonous look and give a soothing effect with their beautiful flowers and foliage. Unsightly, building portions can also be covered and camouflaged by evergreen climbers. They can also be used for training on artistic structures, arches, pergolas, etc. Suitable selection from the following list can be made.

S.No.	Botanical Name	Common Name	Flower Colour	Flowering Season
1.	*Adenocalymma alliaceum*	Adenocaly-mma	Mauve	Early winter
2.	*Allamanda cathartica*	Allamanda	Yellow	Summer
3.	*Antigonon leptopus*	Coral creeper	Pink/white	Rainy Season
4.	*Beaumantia grandflora*	—	White	Spring
5.	*Bougainvillea sp. magenta*	Bougainvillea	Red, pink, orange	All the year intermitently
6.	*Clerodendron spiendens*	—	White scarlet	Winter
7.	*Jacquemontia pentantha*	Jacquemontia	Blue	Summer
8.	*Jasminum grandiflorum*	Chambeli	White	Summer
9.	*Jasminum primulinum*	Peeli Chambeli	Yellow	Summer
10.	*Jasminum Sambac*	Motia	White	Summer
11.	*Passiflora racemosa*	Passion flower	Purple	Rainy season
12.	*Petrea volubilis*	Purple wreath	Pale blue	Spring
13.	*Porana paniculata*	Bridal bouquet	White	Winter
14.	*Pyrostegia vensuta*	Golden shower	Golden orange	Winter
15.	*Quisqualis indica creepter*	Rangoon	White, pink	All the year round in flushes
16.	*Wisteria sinensis*	Wisteria	White or blue	Spring

13

PRESERVATION OF FRUITS AND VEGETABLES

Tomato Juice and Ketchup

Take fully ripe tomatoes having deep-red colour, wash, cut, macerate and boil. Extract the juice in a screw press or household juice extractor. Juice can be taken out with the help of muslin cloth or stainless steel sieve. Discard the seeds and stain. In one kg juice add about 6g salt, 40g sugar and 0.2 sodium benzoate and mix them well. The juice is preserved by filling hot into clean bottles leaving about one inch head space. Close with crown corks and sterlize.

To prepare ketchup take tomato juice/pulp (2.5 kg), onion chopped (40g), garlic chopped (2g), cardamom, black pepper and cumin (0.5g each), cinnamon (1.5g), sugar (125g), salt (30g), red chillies (1g), acetic acid (10ml). Add 1/3rd of sugar to the juice Tie all the spices in a muslin cloth and add to the juice. Boil the juice to concentrate to about 1/3 of its original volume. Remove the muslin bag and squeeze it well to extract the aroma and flavour of the spices. Then add acetic acid, salt and remaining sugar. Test consistency of ketchup by pouring a little of the sample on a dry-plate. If water separate out, cook for more time. Add the perservative sodium benzoate at the rate of 0.75g/kg of the finished product after dissolving sodium benzoate in a small quantity of water. Hot fill and cork.

Take juice/pulp 1.00 kg, sugar 1.75 kg, citric acid 40g, essence 6 ml, preservative 2.8g, water 1.25 l. Add fruit juice/pulp to prepared syrup. Add fruit essence and colour to the desired quantity. Then add preservative in a small quantity of juice. Mix all the ingredients throughly. Fill and cork.

Fruit Jams

For preparation of jam, select slightly under-ripe fruits. Cut the peeled fruits into small bits and wash them. If the fruits are sour add more sugar (1:1) compared to 1:075 in the case of sweet fruits. Citric acid (1.5-2.5g/kg of sugar) is added, depending upon the sweetness of fruits. The mixed contents of the fruits and sugar are cooked to thick consistency until the end-point is reached. The hot jam is poured into clean dry glsss jars and cooled before being closed with screw lids.

Mango Chutney

Fully developed under-ripe mango having yellow pulp can be converted into mango *chutney*. Using ingredients as: Grated peeled mangoes 1kg, sugar 1kg, salt 30g, cardamon 10g, cinnamon 10g, red chilly powder 10g, cumin 10g, onion 25g, ginger 10g, acetic acid 10ml, sodium benzoate 0.6g. Wash and grate the mangoes to obtain thin slices. Add about 250 ml of water to one kg of grated slices and heat the contents to soften the pulp. Add sugar and salt and keep for about 15 minutes. Then add spices, tied loosely in a piece of muslin cloth. Continue heating with frequent stirring with a ladle, until the contents assume thick consistency. Remove the spice bag and squeeze out as much as possible. Continue heating till a product of desired consistency is obtained and then add acetic acid. Cool to room temperature and pack in air tight jars.

Pickle

Prepare vegetable pickle using ingredients as vegetable slice 1 kg, red chillies 20g, black pepper 10g, mustard 50g, cinnamon 6g, dried dates 20g, tamarind (imli) 20g, ginger 10g, salt 90g, acetic acid 50 ml, jaggery (gur) 60g, mustard oil 100g. Wash the vetables thoroughly and cut them into suitable slices. Cook them in boiling water for 5-6 minutes by holding the slices in a muslin cloth. Drain the water and allow the slices to dry superficially spread over a piece of cloth under shade. Chop ginger, onion, garlic and fry them in a little oil. Soak tamarind and dates in water over-night. Remove the stone from the dates and slice them into small pieces. Prepare tamarind extract also. Except jaggery and mustard oil other ingredients are mixed with cooked slices, which are then filled in jars tightly and exposed to sun for 5-6 days. Then add jaggery in the form of thick-syrup and again place the jars in the sun for 3-4 days and add mustard oil. The pickle will be ready in a week's time.

Preserve (Marabbas)

Preserves are prepared from slightly under-ripe fruits like mangoes, apples, carrots and amla. To prepare preserve, wash and peel the fruits. Prick the fruit slices, wrap in muslin cloth and cook in boiling water until soft. Take sugar equal in weight to that of fruits. Make syrup by boiling in water 3 times the weight of sugar. Add 1 to 2g citric acid per kg of sugar used. Add fruits/ slices to the syrup and cook if the syrup drawn between two fingers is thick enough to give 3-4 threads. Allow the product to cool and pack in sterlized dry glass jars.

Lime/Lemon Squash

Fill the bottle with Sugar
Pour Lime/lemon/galgal juice
Keep in sun and shake daily for 7-8 days till whole sugar is dissolved (No preservative is necessary).

14

NUTS AND DRY FRUITS

The role of nuts and dry fruits in human diet, nutrition and their therapeutic values needs no emphasis at the present stage of scientific development and health consciousness. These fruits have special significance for the Himalayan region where climate is eminently suitable (in selected zones) and communication and transport system are inadequately developed. Moreover, the entire produce has to be transported to long distance places in the absence of markets and appreciable consumption locally. In fact the production of high value, low volume durable as against perishables like fruits and vegetables would be a priority for agro-climatic conditions of this area. Nuts (like walnut almond, pecanut, hazelnuts) and dry fruits (like apricot prune and raisin grapes) fit in this category.

Several sub-regions of the Himalayan region are very well suited for their cultivation and further promotion. Some areas of Jammu & Kashmir, Himachal Pradesh and Uttar Pradesh in North West Himalayan region offer favourable conditions. Walnut, almonds, dry apricots are grown on commercial scale in competition with other fruits in Jammu & Kashmir. Similarly almonds, walnut, raisin grapes dry apricots and prunes (dried plum) have been found doing very well in Kinnaur, Lahaul Spiti districts of H.P. Unfortunatly these have not been tried on an appreciable scale in U.P. hill region and therefore, deserve special focus.

One of the basic requirements of the commercial production of nuts and dry fruits is dry climate or say humidity free atmosphere during ripening stage. Wet season (or high humidity condition) is a major impediment. However, walnut is more tolerant and is growing as scattered trees in medium to high hill zones. Some of these yield over a 100 kilograms of nuts (in shell). Many growers are getting over Rs. 1000 per tree. Such success stories need to be reflected in surrounding areas with high quality planting material raised from outstanding trees. It is high time we earmark trees of outstanding merit in our own areas. Almond can also be tried for marketing as green crop.

Even at this stage the fruit is not easily perishable, is not bulky like apples, pear or peach and can be packed in gunny bags and transported to distant markets. Kernels are delicious at this stage. Almonds are harvested when kernel is fully developed but fruit is not infected by gummosis (as yet). Gummosis is the main hurdle in commercial production of dry almonds in monsoon areas.

Whereas dry regions of this state(s) can be specially focussed for production of nuts, dry fruits, walnuts, almonds (green) to some extent pecanuts can be promoted in other areas also. Important cultivars of walnut, almond, pecanut are already given in chapter 5. It may be ensured that hard shell varieties of almond be preferred in the areas with high humidity during ripening time. Two varieties of Hazelnut i.e. Barcelone and Duchilley can be tried. For apricot (drying) it is important that those with higher sugar content (Brix) are selected. Somc of the drying varieties are Shakarpara, Kaisi, Nari and Charmagz.

SAY IT WITH FLOWERS — A—Z

Apple Blossom means "I've got a preference for you". An apple is a symbol of fruitfulness and was, therefore, popular in divination rites.

Broom Sprig means "I am your humble lover". Sprigs were included in bridal bouquets and were used to decorate churches at weddings.

Carnations are associated with both divine and platonic love. A red carnation is a sign of passion and says "I must see you soon".

Daisies give the message "I love your innocence".

Elderflowers say "I have compassion for you". They were held sacred to the Northern goddess of love, Hulda.

Forget-me-nots showed that love was true and could survive long partings.

Honeysuckle is an emblem of rustic beauty and means "I love you warmly". Because of its habit of embracing trees and walls, people call it "love-bind" or "Woodbine".

Ivy promises "friendship in adversity". Being a clinging plant it is regarded as a potent love charm.

Jasmine's greeting is "You are far too kind". Its sweet perfume has made it a favourite love plant since ancient times.

Kingcups, or march buttercups, carry the meaning "I wish rich enough to marry you". People decorated their houses with yellow kingcups because the flower kept love in the home.

Lilies bear the passionate sentiment "My love knows no bounds". This symbol of purity, chastity and innocence is a favourite flower at weddings.

Myrtle means "I love you" and symbolises married bliss. For this reason, as well as its rich fragrance, it is used in bridal bouquets.

Narcissus insists "You are selfish". A legend tells how Narcissus, a handsome young god, was admiring his reflection in a pond when he fell in and was drowned.

Orange Blossom means "You are chaste" so it is commonly

seen on a bride's head-dress. However, its scent is a powerful aphrodisiac which causes sexual excitement when it's perfume is inhaled.

Pansies say "Think of me. You occupy my thoughts night and day". Their name comes from the French work for thought, *pensee*. Other romantic names include Heartsense and Kiss-Me-Quick.

Queen's Rocket should be given to a flirty sweetheart as it says "You are the Queen of Coquettes".

Red Roses means "I love your dearly", but yellow rose bud implies "You are pure and lovely".

Southernwood or **Lad's Love** says "You are teasing me". A sprig of this bush was believed to keep travellers safe and their loved ones happy.

Tulips, when red ask "Will you marry me?". But yellow tulips only mean "You have beautiful eyes".

Ulex, or gorse, whin or **furze**, means "My love is enduring". Its myriad flowers symbolise fertility and sprigs were used in bridal bouquets.

Violet tell the beloved "I am faithful" and were favourite love tokens. Napoleon sent sweet violets to Josephine on each wedding anniversary.

Wallflowers promise "I will be faithful throughout life". This sweet smelling flower was thought to ensure fertility and to promote love between a man and his wife.

Xanthium has a meaning to suit its prickles. Give it to a suitor and it is a sharp reminder that "You are rude and obstinate".

Yarrow means "I kneel at your feet and beg forgiveness". This herb was used to cast love spells.

Zinnias say "I am thinking of absent friends". These colourful flowers come from Mexico where they are associated with platonic love and friendship.

15

LANDSCAPE GARDENING AND FLORICULTURE

Flowers and gardens are an integral part of our customs and culture. Ancient India gardens of Hindu and Buddhist period date back to 2000 to 3000 years B.C. Moghal gardens came into being during 16th and 17th centuries A.D. No religious ceremony is complete (for vast majority of the people) without flowers, fruits and parts of the trees.

India has now some of the world famous gardens which include Rashtrapati Bhawan gardens (Delhi), Brindavan gardens (Mysore), Rose garden (Chandigarh), Lal Bagh (Bangalore), Taj gardens (Agra), Pinjore garden (Pinjore) to name a few. India has also creditable record in the international gardening. India won two gold medals by standing first in outdoor and indoor gardens at the international garden competition at Liverpool UK in 1984. The competition and judging continued for 6 months (after 15 days intervals) and sum total of marks in both categories made India top in both the classes. Thirty-six well-known countries leading in Horticulture particularly gardening participated.

Formal landscape gardening in difficult in hills where topography is broken and area limited. But this broken topography can also be taken advantage of by planning gardens on the pattern of the local topography as 'Nature' would have liked. Uttrakhand has large places of pilgrimage where lakhs of people go every year. Such places need to be given special attention. Infact this will be in tune with our ancient culture when gardens were laid around temples and hermitages of saints.

The hills have wide range of attractive flora and in some

places, beautiful and auspicious myriads of flowers are seen in reasonably large areas, for instance Valley of Flowers in Uttrakhand. This itself is a unique site. Then there are wide stretches of Rhododendrons spread all over, the hills from about 2000m above to 5000m above sea level. While in the lower elevations these are mixed with other flora, in some higher elevation Rhododendrons can be seen in wide stretches. Their flowering is indescribably beautiful and attractive. In these areas, people are dependent on Rhododendrons in many ways. The flower petals are used for making chutney and drinks, leaves are used for spreading under animals and wood as source of energy. It is known to have high medicinal and therapeutic values including antidote for high blood pressure and some enterprising people have started use of petal juice for making of squash on commercial scale. There is immense scope for this venture without much danger to its further proliferation, ecology and environment. Both the tree and man can co-exist for long, if care is taken not to over exploit it.

Flowers

UP hills (as indicated) earlier abound in wild flowers as if nature had created it as a home of flowers. Many people have taken up floriculture as a hobby and for commercial purposes. Uttrakhand can produce off season cut flowers which command a premium in the plain markets. These include Gladiolus, Narcisus, Daffodil and Tuberose. Even roses will produce longer stem in cooler areas. For the time being Gladiolus has caught the imagination of people. The varietal wealth has been added many folds due to liberalisation of imports. A large number of varieties have been imported from Holland. Some of the existing good varieties are: Melody, Snow princess, Sylvia, Suchitra, Mayur, Friendship, Jacksonville Gold, Happy end, Saucerre, Rose supreme and Yellow Stone. The planting time vary according to height and temperature conditions. In the areas adjoining plains they start planting in July-August. In high hills these are planted in November to March-April.

The time of planting of some of the important bulbons and other plants and seasonal flowers for temperate region and valley areas in given below. It can be adjusted accordingly for other climate zones.

S.No.	Name	Time and Planting Temperature Zone	Valleys Area
1.	Agaparittus (African Lilly)	Nov. to March	
2.	Caunas	April, May	March, July, August
3.	Chrysanthemum	March, April	July, August
4.	Dahlia	April, May	July, Sept.
5.	Hermercallis (Day Lilly)	October, March	July, August
6.	Iris	Nov.-March, April	July, August
7.	Kniphifia (Torch Lilly)	March, April	July, August
8.	Narclscus	Sept.	Sept.
a.	Daffodil	Nov.	Oct.
9.	Tulip	Nov.	May, June
10.	Gladiolus	Nov.-March, April	July-Dec.

Seasonal Flowers

S.No.	Winter Tall	Time and Planting Temperature Zone	Valleys Area
1.	Corn flower		
2.	Sweet sultan		
3.	Larkspur		
4.	Poppy		
5.	African Marigold		
6.	Antnhinum		
7.	Sapondia		
8.	Cosmos		

Medium

9.	Aster		
10.	Carnation		
11.	Sweet william		
12.	Pelurues	Transplanted	Seeds sown
13.	Clarkee	in Sept., Oct.	in Sept.
14.	Californian Poppy	before winter	&
15.	Nemsia	sets in	Transplanted
16.	Gazamia		in Oct.
17.	Salvia		
18.	Wall flower		

Dwarf

19. Arcotites
20. Ageratum
21. Daisy
22. Calendula
23. Caudituft
24. Pheox
25. Verbena
26. Mesembryanthemun
27. Dimorpotheca
28. Nasturtium
29. Pansy

Summer

1.	Kochia	Seeds sown in	Seed sown in
2.	Portulaca	March-April	Feb.-March
3.	Coleus	(spring)	&
4.	Zenia	&	Transplanted in
5.	Sunflower	Transplanted in	April
6.	Garlardie	May	
39.	Gompherna		

16

MEDICINAL PLANTS—A HOPE

India lives in her villages. Lot of its population live in rural area where adequate medical facilities are not available. Since time immemorial people have been using herbs in their surroundings and spices in kitchen for their day-to-day ailments. Majority of these medicinal plants are also used by established indigenous medical systems. More than 250 species of such medicinal plants are being used commonly by Ayurvedic physicians, local practitioners and pharmaceutical industry.

With the increasing cost of treatment through allopathic science and adverse effects of synthetic drugs people all over the world are looking towards medicinal plants for their health needs.

Poor socio-economic status of rural population prompts us to explore this medicinal plants wealth which are available in the country and cost effective, use of medicinal plants in common ailments does not often require resources and technology beyond the villages.

So far medicinal plants have been exploited mostly from their natural habitat. But this has been under severe strain due to deforestation, indiscriminate commercial exploitation and export. The traditional pattern of collection and utilisation by Vaidyas is slowly vanishing.

Hence there is an immediate need for conservation and propagation of medicinal plants to meet the individual needs and market demands. There is a need to design simple technologies to convert these medicinal plants into palatable forms to cater to the rural communities.

The Agricultural and Horticultural Universities should become centres of training of various non-governmental agencies regarding the propagation and collection aspects of medicinal plants. They should also impart knowledge on starting of seed banks and nurseries.

The task of the forest research institutes should be the survey of local medicinal flora with the help of Ethnobotanists and conservation of endangered species.

It is matter of great concern that our country and specially Himalayan region being a potential source of a large number of medicinal and herbal plants very little work has been done on its Agro Technology. This has to be worked out in various Agro-climatic conditions. Keeping in mind with the locally available medicinal plant species we need to work further on their propagation methods and simple conversion techniques and where necessary a potential market. There should be a systematic and regular study on market research so that growers are informed about the changing demand pattern.

APPENDICS

APPENDIX - I

ODE TO HIMALAYAS - 1995

M.L. Dewan
Himalaya Sanrakshan Ashram
Ranichauri - 249199
Tehri Garhwal Distt. (U.P.)
Oct. 2, 1995

O wonderful Himalayas nestling sacred dhams
I bow to you as always on this auspicious day
When one hundred twenty-five years ago
Born father of nation as you know
Bapu taught us a lot
We have unlearnt a lot
quoting from Gandhi about the seven sins
Politics without principle
Wealth without work
Pleasure without conscience
Knowledge without character
Commerce without morality
Science without humanity
Worship without sacrifice
O sacred Himalayas you are witness
That we in India are committing
All these seven sins and more
In ecology environment resource pollution and exploitation
Many things happening in your domain
Being built small number of large dams
While we need large number of small dams
Small manageable units for easy governance and development
Instead large state
Hope and pray peoples movement for Uttrakhand spontaneous
Genuine and successful

But beware when new viable state formed
We do not create apathy but love
Create sustainable development
Promote self reliance not dependence
We have adequate resources in Uttrakhand
Of land water people forests
Let us plan and jointly implement viable actions
Like we have mahila shakti programme
Large-scale fuel fodder and multi-purpose tree plantation
Cared for by people
Trees for Life progressing with nut fruit revolution
Also other fruits
Promoting medicinal plants
All based on sacrifice hardwork not exploitation
All in partnership between people
Voluntary agencies and government
Let there be health and education revolution
Vocationalization of environmental careers
Girls and women take their rightful share
in government and in community development
Task is huge journey hazardous and long
Towards peaceful benevolent sacred Uttrakhand
Let us stick together turning sins into virtues
Gandhi initiated to make us strong
Politics with principles
Wealth with work
Pleasure with conscience
Knowledge with character
Commerce with morality
Science with humanity
Worship with sacrifice
And may I add
Human values with love
Resource conservation with integrity
Entrepreneurship with efficiency
Employment with dignity

O mighty Himalayas we ask for your blessings
Give strength courage and wisdom
To all Himalayan community and
Those ready to serve them partner them
Let us march
In your folds towards
A sustainable society
Caring for ecology and environment

APPENDIX - II

RELATIVE CONVERSION TABLE

Group I	Group II	To Convert: Group I to II Multiply by	Group II to I Multiply by
Length			
Inches	Centimetres	2.54	0.39
Feet	Metre	0.30	3.28
Yard	Metre	0.91	1.09
Miles	Kilometres	1.61	0.62
Area			
Sq. Inches	Sq. Centimetres	6.45	0.16
Sq. Feet	Sq. Metres	0.09	10.77
Sq. Yards	Sq. Metres	0.84	1.20
Sq. Mile	Sq. Kilometres	2.59	0.39
Acre	Hectare (1000 Sq. Metres)	0.40	2.47
Volume			
Cubic Inch	Cubic Centimetre	16.39	0.06
Cubic Feet	Litre	28.30	0.04
Cubic Feet	Gallon	6.24	0.16
Cubic Yards	Cubic Metre	0.75	1.32
Capacity			
Imperial Gallon	Litre	4.55	
Weight			
Pound (AV)	Kilograms	0.45	2.20
Ounce (AV)	Grams	28.35	0.04

Grains	Grams	0.07	15.38
Tola	Grams	11.66	0.09
Seer	Kilogram	0.93	1.07
Maund	Quintal (100 kilograms)	0.37	2.68
Hundredweight	Qunital (i.e. 100 kgs)	0.51	1.97
Ton	Metric tonnes (i.e. 1000 kgs)	1.02	0.98
Tola	Grains	1.80	0.01
Carat	Milligram	2.00	0.05
Carat	Tola	0.17	6.00

APPENDIX - III

EQUIVALENTS IN PREPARING CHEMICAL SOLUTIONS

1 Teaspoon	5 ml. or 5 gms
1 Level table spoonful	3 level teaspoonfuls
1 Fluid Ounce	2 table spoonfuls
1 Cupful	8 fluid ounces
1 Pint-2-Cupfuls	16 fluid ounces
1 Quart-2-pints	32 fluid ounces
1 Gallon-4-Quarts	128 fluid ounces
1 Millilitre	1 cubic centimetre
1 Litre-1000 millilitre	1.057 liquid quart
1 Percent-10,000 parts per million	8 fluid ounces 100 gals
1 Pound avoirdupois	453.59 grams
1 Acre	43560 squate feet

APPENDIX - IV

EARTH

The earth is the Lord's and
the fulness thereof,
the world and those who
dwell therein

—Psalm 24.1

And the Lord will make
you abound in prosperity, in
the fruit of your body, and
in the fruit of your cattle,
and in the fruit of your ground....
And the Lord will make you the
head, and not the tail; and you shall
tend upward only, and not
downward; if you obey the
commandments of the Lord
your God.

—Dauteronony-28-11-13

Earth is here so kind, that
just tickle her with a hoe
and she laughs with a
harvest.

—Douglas Jerrold

Earth, which has many
heights and slopes and the
unconfined plain that bind
men together. Earth that
bears plants of various
healing powers,

may she spread wide for us
and thrive.
Earth, in with lie the sea,
the river, and other waters,
in which live all that
breathes and that moves,
may she confer on us the
finest of her yield.

—Atharva Veda 12.1

The earth is garden,
The Lord is gardner,
Cherishing all, none neglected
Adi Graunth Majh Ashtpadi

M3 P 11 8